D0207576

⋙ Landmark Experiments
in Twentieth Century Physics

Landmark
Experiments
in Twentieth Century
Physics

George L. Trigg

Crane, Russak & Company, Inc. • *New York*

Edward Arnold, Ltd. • *London*

Landmark Experiments
in Twentieth Century Physics

Published in the United States by
Crane, Russak and Company, Inc.
347 Madison Avenue
New York, N.Y. 10017

First published in Great Britain in 1975 by
Edward Arnold (Publishers) Ltd.
25 Hill Street
London, W1X 8LL

LC:74-21664

Crane, Russak hard cover ISBN: 0-8448-0602-1
Crane, Russak soft cover ISBN: 0-8448-0603-x
Edward Arnold hard cover ISBN: 0-7131-2506-3

Printed in the United States of America

Contents

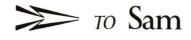 *TO* Sam

who thought it should be written

Preface

Physics is commonly represented as an experimental science. This is certainly valid from the viewpoint of the practicing physicist, who is constantly immersed in the world of current experiment; it is fairly clear to the graduate student in physics, especially the more advanced, who is beginning to immerse himself in that world. The student at the lower level, however, gets very little of the sense of it. On the rare occasions when he is told of a specific experiment, it is usually in a few sentences that give little hint of what really was involved in carrying out the experiment.

To a large extent, the situation is unavoidable, and even proper. It would be impractical to describe in significant detail all the specific experiments the results of which form the fabric of our science. But there is need for some means of giving the student a feel for the nature of experimental work. A useful avenue of approach seemed to be through what I have called "landmark" experiments, investigations that have signaled a marked change either in our ideas about some aspect of nature or in our capability of learning more. This book is the result.

The choice of experiments to include is, of course, essentially arbitrary. I decided to restrict myself to the twentieth century and to omit these areas—relativity, purely nuclear physics, and early atomic physics—that have been covered in some detail elsewhere. But even within these limits, others would doubtless make somewhat different choices. I offer no apologies for mine.

I have attempted to present each of the experiments in the framework of the physics of its time, to give a feeling for why it was done at the time it was, and how. However, while I endeavored to make no historically inaccurate statements, I make no pretense of being a historian or of writing a history of any experiment.

I have benefited greatly from the fact that many of the people whose work is described were still alive while my work was in progress, and that I am acquainted with many of them. I wish to acknowledge valuable conversation and/or correspondence with the following: J. F. Allen, P. W. Anderson, John Bardeen, Martin M. Block, Walter H. Brattain, the late Clyde L. Cowan, Jr., Russell J. Donnelly, Ivar Giaever, S. A. Goudsmit, Lee Grodzins, Martin Klein, Nicholas Kurti, P. Kusch, Willis E. Lamb, Jr., Leon M. Lederman, James E. Mercereau, Rudolf Mössbauer, A. A. Penzias, Fred Reines, A. H. Rosenfeld, J. M. Rowell, N. P. Samios, A. L. Schawlow, J. R. Schrieffer, Sidney Shapiro, William Shockley, C. A. Swenson, Barry N. Taylor, V. L. Telegdi, C. H. Townes, Charles Weiner, and C. S. Wu. To those in this list whose work is treated in the book, I give particular thanks for their permission to use their reports. I obtained helpful material on the development of the transistor from the Niels Bohr Library of the Center for the History of Physics, at the American Institute of Physics. I appreciate the cooperation of the staff of the Research Library at Brookhaven National Laboratory, and of the Kline Library of Science at Yale University. I had valuable procedural assistance from Bob Adair, Jack Greenberg, and Pete Martin. I thank Leslie Freeborn, Debbie Ierardi, Connie LoRe, Chris Saïtta, Kit Sergio, and Carolyn Wuest for reducing a sometimes confusing manuscript to an orderly typed version. Finally, I am grateful to D. A. Bromley and the Department of Physics at Yale University for hospitality during a sabbatical leave early in 1972, when this work was started.

George L. Trigg

Acknowledgments

I am indebted to the following sources for permission:

To Johannes Ambrosius Barth Verlag, for translated quotations and figures from a paper by Friedrich, Knipping, and Laue, from *Annalen der Physik*.

To Taylor and Francis Ltd., for quotations and figures from papers by Thomson and by Moseley, from *Philosophical Magazine*.

To the Royal Society, for quotations and figures from papers by Thomson, by Daunt and Mendelssohn, and by Allen and Misener, from the *Proceedings of the Royal Society*, *Series A*.

To the Royal Institution of Great Britain, for a quotation and a figure from a paper by Thomson, from *Proceedings of the Royal Institution*.

To Eduard IJdo N V, for quotations and figures from papers by Kamerlingh Onnes, by Kamerlingh Onnes and Boks, by Keesom, and by Keesom and Clusius, from *Communications of the Physical Laboratory of the University of Leiden*.

To S. Hirzel Verlag, for translated quotations and figures from a paper by Meissner, from *Physikalische Zeitschrift*.

To North-Holland Publishing Company, for quotations and figures from papers by Keesom and Miss Keesom, from *Physica*.

To Macmillan and Company Ltd., for quotations and figures from papers by Kapitza, by Allen and Misener, by Allen and Jones, by Daunt and Mendelssohn, by Kikoin and Lasarew, and by Maiman, from *Nature*.

To The American Physical Society, for quotations and figures from papers by Scott; by Rabi, Zacharias, Millman, and Kusch; by

Millman, Rabi, and Zacharias; by Lamb and Retherford; by Kusch and Foley; by Bardeen; by Bardeen and Brattain; by Wu, Ambler, Hayward, Hoppes, and Hudson; by Garwin, Lederman, and Weinrich; by Friedman and Telegdi; by Reines, Cowan, Harrison, McGuire, and Kruse; by Gordon, Zeiger, and Townes; by Gordon; by Schawlow and Townes; by Maiman; by Maiman, Hoskins, D'Haenens, Asawa, and Evtuhov; by Giaever and Megerle; by Rowell and Anderson; by Rowell; by Shapiro; by Giaever; by Langenberg, Scalapino, Taylor, and Eck; by Jaklevic, Lambe, Silver, and Mercereau; by Parker, Langenberg, Denenstein, and Taylor; by Barnes et al.; and by Woody et al., from *The Physical Review* and *Physical Review Letters*.

To Elsevier Scientific Publishers, for quotations and figures from the Nobel lecture of Bardeen, from *Nobel Lectures in Physics*.

To the American Association of Physics Teachers, for quotations and a figure from a paper by Brattain, from *The Physics Teacher*.

To McGraw-Hill Book Company, for a quotation from *Physics for Poets*, by R. H. March.

To Springer-Verlag, for translated quotations and figures from a paper by Mössbauer, from *Zeitschrift für Physik*.

To *Zeitschrift für Naturforschung*, for translated quotations and figures from a paper by Mössbauer.

To the Smithsonian Institution, for a quotation and figures from a paper by Cowan, from the *Smithsonian Report for 1964*.

To American Telephone and Telegraph Company, for quotations and figures from a paper by Crawford, Hogg, and Hunt, from *Bell System Technical Journal*.

To the American Institute of Physics, for a quotation from a paper by Anderson, from *Physics Today*, and a quotation and a figure from a paper by Penzias, from *Reviews of Scientific Instruments*.

To the University of Chicago Press, for quotations and a figure from papers by Penzias and Wilson, from *Astrophysical Journal*.

The following frequently cited sources are quoted in abbreviated form:

Nobel Prize Lectures—Physics (Elsevier Scientific Publishing Company, New York, 1966, 1965, and 1964), cited as *Nobel Lectures*.

World of the Atom, edited by Henry A. Boorse and Lloyd Motz (Basic Books, New York, 1966), cited as *World of the Atom*.

George L. Trigg, *Crucial Experiments in Modern Physics* (Van Nostrand Reinhold, New York, 1971 and Crane, Russak, & Co., New York, 1975), cited as *Crucial Experiments*.

Chapter 1

The Wave Nature
of X Rays

Scarcely any other invention in history was exploited as promptly as were x rays. Within a few months of their discovery by Wilhelm Röntgen in 1895, they were being used for medical diagnostic purposes and for the examination of metal castings. Yet it was not until nearly twenty years later that their true nature was established.

The question of the nature of x rays had been widely argued almost from the start. Many scientific observers, as well as a large part of the general public, appeared to regard them as identical to cathode rays, the beams of electrons emitted from the cathode of an electrical discharge in a low-pressure gas, despite the fact that they were unaffected by magnetic fields. Other scientists thought that they were longitudinal vibrations in the "aether"; still others suspected that they were transverse waves of a character similar to light. The difficulty lay in the fact that the known properties and producible effects did not seem to fit any of these hypotheses. When the rays struck matter, they were scattered, much as light is scattered from a cloudy liquid. But they could not be refracted or reflected.[1] Efforts to produce polarization by selective absorption, in the manner used for visible light in tourmaline, were also unsuccessful. Charles Barkla, in 1906, demonstrated polarization by double scattering,[2] but many people were not convinced, since such experiments could also be explained in terms of spinning particles. The real touchstone of the wave nature, as Thomas Young had recognized in regard to visible light a century earlier, was the production of interference effects. Attempts in this direction were hampered by a lack of knowledge of the wavelength range. The decisive work was done in 1912 by Max von Laue, Walter Friedrich, and Paul Knipping, and earned a Nobel prize in physics for Laue in 1914. This chapter describes their work, as originally presented to the Royal Bavarian Academy of Sciences and published in its Meeting Reports, and subsequently republished in *Annalen der Physik*.

Actually, previous efforts to detect interference had been made. As early as 1899, Hermann Haga and Cornelis Wind had passed a beam of x rays through a triangular aperture. If the x rays were waves, they should have been diffracted by the edges of the slit, and the image on a photographic plate some distance behind the slit should have been broader than the slit itself[3]; the amount of broadening, together with the dimensions of the apparatus, would then give an estimate of the wavelength. Haga and Wind concluded that if there were interference effects, the wavelengths involved must be less than about 10^{-9} cm. This work was repeated by Bernhard Walter and Robert Pohl in 1908, with even more discouraging results—they put limits of the order of 10^{10} cm on the wavelength. Their work, however, was reanalyzed in 1912 by Arnold Sommerfeld, with the help of photometric measurements on the original photoplates by Peter Koch, and the conclusion of Haga and Wind was supported: No assurance could be found that waves were actually involved, but if they were, they must have wavelengths of the order of 10^{-9} cm.

What Laue did was to fit this piece of data with others from the theory of solids and atomic theory. He knew, first, that "already in 1850 there was introduced into crystallography by Bravais the theory that the atoms in crystals are arranged in a spatial lattice. If the Röntgen rays truly consist of electromagnetic waves, then it is to be expected that on excitation of the atoms to vibrations, free or forced, the space-lattice structure will give rise to interference phenomena." Moreover, "the constants of this lattice can be easily calculated from the molecular weight of the crystallized compound, its density, and the number of molecules per gram molecule, in addition to the crystallographic data. One finds for them just the order of magnitude 10^{-8} cm. . . ." This was just what was needed to produce significant interference phenomena with x rays, if indeed the rays were wavelike in character.

The known results of optical interference theory could not be taken over directly, because of the "considerable complication . . . that for the space lattice a threefold periodicity is present, whereas for optical gratings one has a periodic repetition only in one direction, or . . . at most two directions." Laue, therefore, worked out the theory on the basis that each atom was excited an equal amount by the influence of an incident plane wave traveling at the speed of light. The details of this derivation are of no concern here; the crucial result was a set of three equations for the direction in which the scattered intensity would have a maximum. Each of these equations "represents a set of circular cones whose axes coincide with one of the edges" of the elementary unit of the space lattice. "Now, obviously, only in exceptional cases will it happen that one direction satisfies all three conditions at the same time.... Nevertheless, a visible maximum of intensity is to be expected when the line of intersection of two cones of the first two sets lies close to a cone of the third set." If the scattered rays strike plane photographic plates, these maxima will produce isolated spots, which, however, should be grouped along families of conic sections[4]—more particularly, at points where three curves, one from each of three families, intersect or nearly intersect.

"It must be observed that for a given space lattice the division into elementary parallelepipeds is not unique, but can be carried out in innumerable ways.... According to the foregoing, the intensity maxima must be able to be grouped along interrupted conic sections around ... axes [arising from such alternative divisions], as in general, to each such kind of division belongs a way of grouping the maxima."

At Laue's suggestion, Friedrich and Knipping carried out the experimental test. "After some preliminary studies with a provisional apparatus," the apparatus shown in Fig. 1-1 was built. "From the Röntgen rays proceeding from the anticathode A of a Röntgen tube, a small pencil of about 1 mm diameter was cut off by the stops B_1 to B_4. This pencil penetrates the crystal Kr, which is set up on a goniometer G. Around the crystal, in various directions and at different distances, were fixed photographic plates P, on which was recorded the intensity distribution of the secondary radiation emanating from the crystal. The setup was guarded against unwanted radiation in a satisfactory way by a large lead shield S as well as by the lead case K.

"The arrangement of the entire experimental setup was effected by optical means. We had a cathetometer, whose telescope was fitted with a crosshair, set up immovably. The 'hot spot' of the anticathode, the stops, and the goniometer axis were brought in turn into the optical axis of the telescope. ... The stops B_1 to B_3 mainly blocked off the secondary radiation from the tube walls, while stop B_4 formed the limits for the pencil of Röntgen rays incident on the crystal. This ordinarily had a diameter 0.75 mm, was drilled in a lead disk 10 mm thick, and could be adjusted by means of three

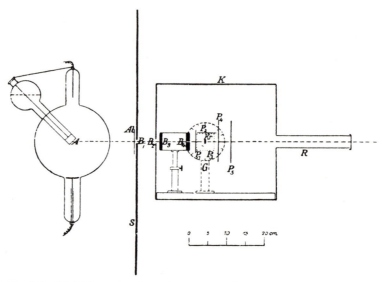

Fig. 1-1. Friedrich and Knipping's apparatus for studying the scattering of x rays penetrating a crystal. [*Ann. Physik* **41** (1912), p. 979, Fig. 1.]

positioning screws (not shown) in such a way that the axis of the hole coincided exactly with the axis of the telescope or the axis of the pencil of rays. In this way it was arranged that a ray pencil of circular cross section fell on the crystal. . . . The tube R served . . . to avoid as much as possible the secondary rays that were produced by the primary radiation striking the rear wall of the case.

"After this adjustment, . . . the axis of the goniometer was set perpendicular to the path of the rays in the usual way. In the same way the different plate holders were . . . adjusted. . . . When the apparatus was oriented to that extent, the crystal to be irradiated, which was fastened to the goniometer table by a trace of sticky wax, was put in place, this again with the help of the telescope already mentioned. . . . This . . . very essential adjustment we could make to within a precision of a minute [of arc]."

The first exposure with the final apparatus used a "middling" crystal of copper sulfate that had been used in the preliminary studies. The figures obtained on plates P_4 and P_5 in this exposure are shown in Fig. 1-2. "It is noteworthy that the distances (crystal–P_4) and (crystal–P_5) are in proportion to the sizes of the figures on P_4 and P_5, respectively, from which it is established that the rays travel out from the crystal in straight lines. It is further to be observed that the sizes of the individual secondary spots, despite the greater distance of plate P_5 from the crystal, remain the same.

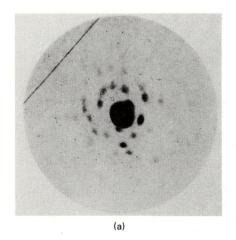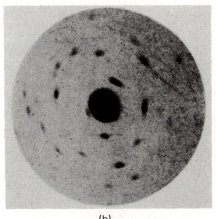

(a)   (b)

Fig. 1-2. Two figures obtained in the first exposure: (a) from plate P_4 of Fig. 1-1, (b) from plate P_5. [*Ann. Physik* **41** (1912), Plate I, Figs. 2 and 1, respectively.]

This is taken as a sure indication that the secondary rays producing each individual spot leave the crystal as a parallel beam.

"It is to be expected that the phenomenon will be more transparent and easy to understand with a crystal of the regular [i.e., cubic] system than with the triclinic copper sulfate, since it can be assumed[5] with certainty that the pertinent space lattice is of the greatest possible simplicity. Regular zinc blende seemed suitable to us. . . . We had a plane parallel plate of 10×10 mm dimensions and 0.5 mm thick cut . . . parallel to a cube face (perpendicular to a crystallographic principal axis) from a good crystal. This plate was oriented . . . so that the primary rays penetrated the crystal perpendicular to the cube face. Figure [1-3] shows the result of one such trial. The pattern of the secondary spots is completely symmetric around the position of the unscattered beam. . . . [The fourfold nature of the symmetry] is certainly one of the most beautiful pieces of evidence for the space lattice of the crystal, and that no property other than the space lattice alone comes into play here."

A slightly later paper by Laue contained a more detailed analysis. Laue had, as noted earlier, already derived the equations for the directions of the maxima. For the case of a cubic crystal with the incident beam directed along one of the principal axes, they take the simple form

$$\alpha = h_1 \lambda/a, \qquad \beta = h_2 \lambda/a, \qquad 1 - \gamma = h_3 \lambda/a, \qquad (1\text{-}1)$$

where λ is the wavelength; a is the length of one edge of the elementary cubic unit of the crystal lattice; α, β, and γ are the cosines

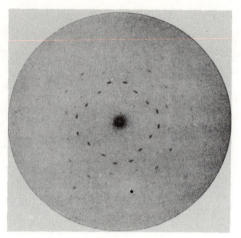

Fig. 1-3. The pattern of spots produced by x rays after passing through a crystal of zinc sulfide. [*Ann. Physik* **41** (1912), Plate II, Fig. 5.]

of the angles between the direction of the maximum and the x, y, and z axes, respectively (the z axis being the direction of the incident beam); and h_1, h_2, and h_3 are integers (positive, negative, or zero). He was able to account for all the points in Fig. 1-3 by suitable choices of the three h's and the assumption that the radiation consisted of five discrete wavelengths.

Laue's analysis was soon subjected to criticism by W. L. Bragg, who noted that there were several sets of h's which would satisfy the three components in Eq. (1-1) for one or another of the five wavelengths, and yet for which there was no spot. Bragg proposed an alternative explanation: that the incident radiation contained a continuous distribution of wavelengths, and that the maxima were produced by interference, not of beams from individual atoms, but of beams reflected from families of parallel planes. This mechanism will produce maxima whenever the wavelength λ, the angle θ between the incident beam and the normal to the planes, and the spacing d between planes in a family are related by

$$n\lambda = 2d \cos \theta,$$

where n is a positive integer. Bragg took $n = 1$ throughout. The discreteness of the spots results from the fact that in a crystal lattice, only discrete families of planes of atoms can be formed. Bragg's analysis proved much more satisfactory than Laue's, in that it accounted not only for the positions of the spots but also qualitatively for their intensities; and Bragg and his father, Sir W. H. Bragg, soon applied it to further studies of x rays and of crystals.

Indeed, Laue had said in the first paper that agreement of his computation with experiment did "not hide the fact that our theory can be improved in every respect." The essential conclusions, however, were qualitative ones that were not altered by details of the analysis. "That the radiation leaving the crystal has a wave character is clearly shown by the sharpness of the intensity maxima, which is easily understandable as an interference phenomenon, but scarcely so on the basis of corpuscular concepts. . . . Nevertheless one could perhaps doubt the wave nature of the primary rays. Let us suppose that the atoms of the crystal were . . . excited by a corpuscular radiation. . . . In that case there could be set into coherent oscillations only those rows of atoms that were struck by the same particle. . . . Consequently . . . we would obtain only one condition for an intensity maximum and this, as already evident on grounds of symmetry, would be fulfilled on a circle around the point of impact of the primary rays. The interruption of this circle, which indeed actually appeared, would then not be understandable. But besides, the primary rays and those emanating from the crystal are to all appearances so similar that one can with passable certainty deduce the wave nature of the latter from that of the former."

Notes

1. Reflection was achieved in 1923. Even then it was not the kind of reflection ordinarily thought of for visible light, but the special type usually referred to as total "internal" reflection; and it took place only at near-grazing incidence.

2. An interpretative description of such an experiment in terms of transverse electromagnetic waves is as follows: Consider an unpolarized beam incident at 45° on a scattering surface. The beam leaving also at 45°, and thus at right angles to the incident, will be at least partially polarized (the degree depending on the optical properties of the scatterer), with the electric vector preferentially parallel to the scattering surface. Consequently, a second scattering through 90° will give a final scattered intensity that will depend strongly on the azimuth of the final beam around the intermediate. In particular, the scattering will be minimum when the final beam is perpendicular to the original beam as well as to the intermediate, and maximum when the final beam is parallel or antiparallel to the original.

3. The reason for use of a triangular slit is twofold: first, the aforementioned lack of knowledge of the wavelength and thus of a satisfactory width for a straight slit; second, the fact that the broadening would in any case be greater, as well as easier to establish, near the apex end than near the base end.

4. A conic section is a curve formed by the intersection of a plane and a circular cone. Conic sections include circles, parabolas, ellipses, and hyperbolas.

5. The certain deciphering of crystalline structure became possible only as a consequence of the work being described here.

Bibliography

The original papers are in German: W. Friedrich, P. Knipping, and M. v. Laue, *Sitzungsberichte der königliche Bayerische Akademie der Wissenschaften, Mathematische-Physikalische Abteilung* 1912, 303; M. v. Laue, ibid., 363. A translated abridgement of the first paper appears in *World of the Atom*, vol. I, p. 832. These are reproduced "slightly altered" in W. Friedrich, P. Knipping, and M. v. Laue, *Annalen der Physik* 41, 971 (1913); M. v. Laue, ibid., 989. In the reprinting, however, the numbering of the figures in Plate I was reversed from that in the original and in the texts: See W. Friedrich, P. Knipping, and M. v. Laue, *Annalen der Physik* 42, 1064 (1913).

Laue describes the developments in his Nobel prize lecture, in *Nobel Lectures*, vol. I, p. 347.

Bragg's analysis is presented in W. L. Bragg, *Proceedings of the Cambridge Philosophical Society* 17, 43 (1912). His work and that of his father won them a joint Nobel prize; see his lecture in *Nobel Lectures*, vol. I, p. 370.

An annotated bibliography of further readings pertaining to x rays is given by L. Muldawer, *American Journal of Physics* 37, 123 (1969).

Chapter 2

Isotopes

Electrical discharges in gases have been a subject of considerable attention ever since the middle of the nineteenth century, when the invention of the mercury-displacement pump by Geissler and of the induction coil by Faraday combined to allow production of the discharges under controlled conditions, thereby making such attention meaningful. Evidence of their significance in illumination is all around us; and since about 1950, their importance to the potentiality of power from nuclear fusion has made them the subject of special interest. What may not be so widely recognized, however, is the extent to which they have served as a means through which much of our knowledge of the atom has been obtained. They were used as sources for spectroscopic studies; they provided, through the discovery of the electron, the first unmistakable evidence that the atom is not indivisible; they were the actual object of study when Röntgen discovered x rays; and it was in a study of one of their aspects, the positive rays, that Sir J. J. Thomson found the first clear indication that not all the atoms of a given chemical element are truly identical. This chapter presents the story of that discovery.

In 1886, Eugen Goldstein discovered that if the cathode (negative electrode) in a discharge tube was perforated, streams of luminosity appeared behind it, i.e., on the side away from the discharge. He gave these luminous streams the name *Kanalstrahlen* or *channel rays*, because the perforations were small holes through a comparatively thick plate. However, he was not able to deflect them by means of the magnetic fields available to him. In 1898, Willy Wien succeeded in causing magnetic deflection of the rays. Not only did this establish that they were (positively) charged; it also enabled him to estimate the ratio of their charge to their mass.[1] The value he obtained was of the same order of magnitude as those obtained for the charge carriers in electrolysis.

Thomson regarded the rays as "the most promising subjects for investigating the nature of positive electricity," and therefore undertook a more careful "series of determinations of the values of e/m for positive rays under different conditions." Most of his results were published in a series of eight papers in the *Philosophical Magazine*, extending from 1907 to 1912; they were also presented in part at a meeting of the Cambridge Philosophical Society and three of the weekly meetings of the Royal Institution; and they were summarized in a lecture before the Royal Society and in a book. Despite this seemingly extensive record, the story is not an easy one to unravel. There were, not surprisingly, several stages of development of the research; there were topics that were of considerable interest to Thomson that now appear relatively insignificant; and Thomson's writings are not always completely clear, even when complemented by drawings. The end result, however, is unmistakable.

Although the basic features of the method remained unchanged, the details developed as the work went forward. In all cases, "the object of the experiments was to determine the value of e/m by observing the deflexion produced by magnetic and electric fields." At first, "the rays were detected and their position determined by the phosphorescence they produced on a screen at the end of the discharge-tube. A considerable number of substances were examined" to select the most satisfactory phosphor; likewise, "considerable trouble was found in obtaining a suitable substance to make the powder adhere to the glass. . . .

"The form of the tube adopted is shown in fig. [2-1]. A hole is bored through the cathode, and this hole leads to a very fine tube F. The bore of this tube is made as fine as possible so as to get a small well-defined fluorescent patch on the screen. These tubes were either carefully made glass tubes, or else the hollow thin needles used for hypodermic injections, which I find answer excellently for this purpose. After getting through the needle, the positive rays on their way down the tube pass between two parallel aluminum plates A, A. These plates are vertical, so that when they are maintained at different potentials the rays are subject to a horizontal electric force, which produces a horizontal deflexion of the patch of light on the screen. The part of the tube containing the parallel aluminum plates is narrowed as much as possible, and passes between the poles P, P of a powerful electromagnetic. . . . The poles of this magnet are as close together as the glass tube will permit, and are arranged so that the

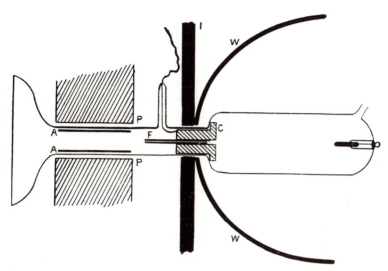

Fig. 2-1. Early version of Thomson's apparatus for measuring *e/m* for positive rays. [*Phil. Mag.* **13** (1907), p. 563, Fig. 2.]

lines of magnetic force are horizontal and at right angles to the path of the rays. The magnetic force produces a vertical deflexion of the patch of phosphorescence on the screen. To bend the positive rays it is necessary to use strong magnetic fields, and if any of the lines of force were to stray into the discharge-tube in front of the cathode, they would distort the discharge in that part of the tube. This distortion might affect the position of the phosphorescent patch on the screen, so that unless we shield the discharge tube we cannot be sure that the displacement of the phosphorescence is entirely due to the electric and magnetic fields acting on the positive rays after they have emerged from behind the cathode.[2]

"To screen off the magnetic field, the tube was placed in a soft iron vessel W with a hole knocked in the bottom, through which the part of the tube behind the cathode was pushed. Behind the vessel a thick plate of soft iron with a hole bored through it was placed [presumably I in Fig. 2-1], and behind this again as many thin plates of soft iron, such as are used for transformers, as there was room for were packed. When this was done it was found that the magnet produced no perceptible effect on the discharge in front of the cathode.

". . . When the rays were undeflected they produced a bright spot on the screen; when the rays passed through electric and magnetic fields the spot was not simply deflected to another place, but was drawn out into bands or patches, sometimes covering a considerable area. To determine the velocity of the rays and the value of e/m, it was necessary to have a record of the shape of these patches. . . . The method actually adopted was as follows:—The tube was placed in a dark room from which all light was carefully excluded, the tube itself being painted over so that no light escaped from it. Under these circumstances the phosphorescence on the screen appeared bright and its boundaries well defined. The observer traced in Indian ink on the outside of the thin flat screen the outline of the phosphorescence. When this had been satisfactorily accomplished the discharge was stopped, the light admitted to the room, and the pattern on the screen transferred to tracing-paper; the deviations were then measured on these tracings.''

In order to interpret the patterns, it is necessary to calculate the effects of the fields. This can be done rather simply for the electric deflection, and approximately for the magnetic. The necessary geometry is shown in Fig. 2-2. Consider first the case of the electric field,

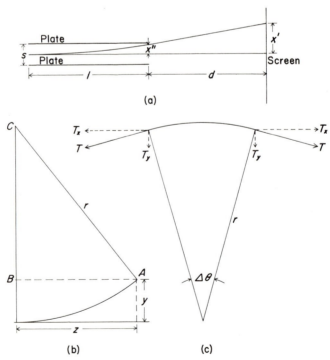

Fig. 2-2. Geometrical considerations for calculating deflections of the positive rays: (a) case of the electric field; (b) case of the uniform magnetic field; (c) case of the wire under tension.

as shown in part (a). It will be assumed that the field is uniform between the plates and zero elsewhere (Thomson in the sixth paper of the series carried through the analysis including edge effects, and got results not significantly different). If the separation of the plates is s and the potential difference between them is V, the value of the field strength is V/s, and a particle of charge e and mass m will experience a constant acceleration of $(e/m)V/s$ in the x direction. If its initial velocity is v, it will spend a time l/v in the field during which it will acquire a transverse velocity

$$v_x = a \frac{l}{v} = \frac{e}{m} \frac{V}{s} \frac{l}{v},$$

and it will have undergone a deflection

$$x'' = \frac{1}{2} a \left(\frac{l}{v}\right)^2 = \frac{1}{2} \frac{e}{m} \frac{V}{s} \left(\frac{l}{v}\right)^2$$

It will take a time d/v to travel the remaining distance d to the screen, so that its additional displacement in the x direction will be $v_x(d/v) = eVld/msv^2$. The total deflection at the screen is this plus x'', or

$$x' = \frac{1}{2} \frac{e}{mv^2} \frac{V}{s} (2ld + l^2) = \frac{1}{2} \frac{e}{mv^2} \frac{V}{s} l(l + 2d). \qquad (2\text{-}1)$$

The magnetic deflection is also reasonably simple for an idealized situation in which there is a uniform field extending all the way to the screen. In this case, the path is a circular arc; the magnetic force Bev, which always acts at right angles to the direction of motion, provides the centripetal force mv^2/r:

$$Bev = mv^2/r,$$

from which the reciprocal of the radius is given as

$$1/r = Be/mv. \qquad (2\text{-}2)$$

Now by application of Pythagoras's theorem to the triangle ABC in part (b) of Fig. 2-2, there is obtained

$$r^2 = (r - y)^2 + z^2;$$

expansion of the square and transposition gives

$$2ry - y^2 = z^2.$$

For the usual situation, the deflection y is much smaller than the radius of curvature r, so that the y^2 can be neglected; then the solution for y is

$$y = \frac{1}{2} \frac{z^2}{r}. \tag{2-3}$$

Substitution of the above value for $1/r$ gives

$$y = \frac{1}{2} \frac{e}{mv} Bz^2. \tag{2-4}$$

This holds for the value of y at any corresponding value of z; in particular, the magnetic deflection at the screen in this ideal case would be

$$y' = \frac{e}{mv} B(l + d)^2. \tag{2-5}$$

In actuality, however, the field is not uniform; Eq. (2-4) is replaced by

$$y = \frac{e}{mv} B_0 f(z), \tag{2-4a}$$

and Eq. (2-5) by

$$y' = \frac{e}{mv} B_0 f(l + d), \tag{2-5a}$$

where B_0 is the value of the field at some arbitrary reference point and $f(z)$ is a function that depends on how the field varies from point to point.[3]   This function $f(z)$ cannot be computed and must therefore be determined empirically; fortunately, only its value for $z = l + d$ is needed. The method of determining it depends on the fact that a flexible wire carrying a current in a magnetic field will tend to form itself into a curve whose radius of curvature at any point is given by

$$1/r = Bi/T, \tag{2-6}$$

where i is the current in the wire and T is the tension in the wire. [To see this, consider part (c) of Fig. 2-2. The magnetic force on the segment of wire is $Bir\Delta\theta$, since the length of the segment is $r\Delta\theta$. This force is radially outward and must be balanced by the radially inward components of the tension forces T, T, each of which is of magnitude $T \sin (\frac{1}{2}\Delta\theta)$. Thus $2T \sin (\frac{1}{2}\Delta\theta) = Bir\Delta\theta$. For very small $\Delta\theta$, it is a good approximation to replace $\sin (\frac{1}{2}\Delta\theta)$ by $\frac{1}{2}\Delta\theta$, and Eq. (2-6) follows immediately.] This is exactly analogous to Eq. (2-2). Consequently, if such a wire is placed in the field and its outer end is adjusted vertically until its inner end is horizontal so as to match the conditions of the beam of positive rays, it will position itself along the curve given by

$$y = (i/T)B_0 f(z),$$

and its outer end will be at a distance y_1 from the horizontal axis given by

$$y_1 = (i/T)B_0 f(l + d). \tag{2-7}$$

Dividing Eq. (2-7) into Eq. (2-5a) gives

$$\frac{y}{y_1} = \frac{e/mv}{i/T},$$

from which not only the unknown function $f(l + d)$ but also the reference value B_0 has been eliminated. Actually, to allow the use of the relations involved without having to obtain one or another of y_1, i, or T for every value of B_0 used, it is simpler to leave in the value B_{0w} used with the wire and write

$$\frac{y_1}{y} = \frac{e/mv}{iB_{0w}/T} B_0,$$

or

$$y = \frac{e}{mv} \frac{y_1 T}{iB_{0w}} B_0. \tag{2-8}$$

The application of the method is shown in Fig. 2-3. "The part of the tube through which the rays pass was cut off, and a metal rod placed so that its tip Z coincided with the aperture of the narrow

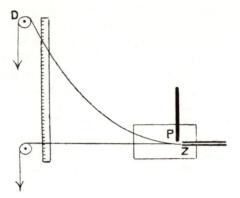

Fig. 2-3. Use of the flexible wire to determine the magnetic field distribution. [*Phil. Mag.* **13** (1907), p. 565, Fig. 3.]

tube through which the positive rays had emerged. A very fine wire soldered to the end of this tube passed over a light pulley, and carried a weight at the free end. The pulley was supported by a screw by means of which it could be raised or lowered; a known current passed through the wire, entering it at Z and leaving it through the pulley. The pulley was first placed so that the path of the stretched wire when undeflected by a magnetic field coincided with the path of the undeflected rays. A vertical scale whose edge was the same distance from the opening through which the rays emerge as the screen on which the phosphorescence had been observed, was placed just behind the wire, and was read by a reading microscope with a micrometer eyepiece. . . . To ensure that the tangent to the wire is horizontal when $z = 0$, the following method is used. P is a chisel-edge carried by a screw and placed about 1 mm in front of the fixed end of the wire; this is adjusted so that when the magnetic field is not on the wire just touches the edge: this can be ascertained by making the contact with the wire complete an electric circuit in which a bell is placed. When the magnetic field is turned on the wire is pulled off from the edge, and the tangent at $z = 0$ is no longer horizontal; it can, however, be brought horizontal by raising or lowering the pulley D until the wire is again in contact with P, which can be ascertained again by the ringing of the bell. Then y_1 is the vertical distance between the point where the wire now crosses the edge of the scale and the point where it crossed before the magnetic field was put on."

Returning now to Eqs. (2-1) and (2-8), it can be seen that they contain two parameters of the positive rays: the charge-to-mass ratio e/m, and the initial speed v. For the moment, let them be written in the forms

$$x' = C_1 e/mv^2,$$

$$y' = C_2 e/mv,$$

where C_1 and C_2 are constants depending on the electric and magnetic fields and the dimensions of the apparatus. If e/m is eliminated from these last two (by dividing one by the other), the result is

$$y'/x' = (C_2/C_1)v;$$

if the velocity is eliminated, the result is

$$y'^2/x' = (C_2^2/C_1)e/m. \tag{2-9}$$

"We see [from the first of these] that if the pencil is made up of rays having a constant velocity but having all values of e/m up to a maximum value, the spot of light will be spread out by the electric and magnetic fields into a straight line extending a finite distance from the origin. While if it is made up of two sets of rays, one having the velocity v_1 the other the velocity v_2, the spot will be drawn out into two straight lines. . . .

"If e/m is constant and the velocities have all values up to a maximum, the spot of light will be spread out into a portion of a parabola. . . .

"The discharge was produced by means of a large induction-coil, giving a spark of about 50 cm in air, with a vibrating make and break apparatus. Many tubes were used in the course of the investigation, the dimensions of these varied slightly."

The first results were rather unpromising. The sizes of the tubes used were such that the discharge could not be maintained at very low pressures without sparking through the glass envelope. A typical pattern produced in air at a pressure of order 1/50 mm Hg is shown in Fig. 2-4. "The deflexion under magnetic force alone is indicated by vertical shading, under electric force alone by horizontal shading, and under the two combined by cross shading.

"The spot of phosphorescence is drawn out into a band on either side of its original position. The upper portion . . . indicates that the phosphorescence is produced by rays having a positive charge; the lower portion (indicated by dots in the figure) . . . is deflected as if *the rays carried a negative charge.* . . . Considering for the present the upper portion, the straightness of the edges shows that the velocity of the rays is approximately constant, while the values of

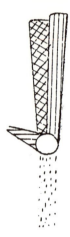

Fig. 2-4. A typical "high"-pressure pattern of phosphorescence. [*Phil. Mag.* **13** (1907), p. 568, Fig. 6.]

e/m range from zero at the undeflected portion to the value approximately equal to 10^4 at the upper end."

By use of special electrode materials, Thomson was able to obtain some results at substantially lower pressures; an example taken from a discharge in helium is shown in Fig. 2-5. These results were sufficiently interesting to spur him on to further efforts to work at low pressures. He succeeded in doing so, as reported in his sixth paper, by working with tubes of much larger volume in the discharge portion. With such a tube, he followed the changes in pattern as the pressure was gradually reduced. At first it was a straight band, like that in Fig. 2-4. "... [T]his persists for a considerable range of pressure, but as the pressure is still further reduced bright spots ... begin to appear, while the luminosity appears to divide into two portions, the appearance being that represented in fig. [2-6].

"The luminous band, which at the higher pressures was the sole representative of the phosphorescent, can still be seen in its old position though it is not so bright as when the pressure was higher, the negative continuation of it persists. As the pressure is still further diminished this part of the phosphorescence with its negative accompaniment gets fainter and fainter ..., finally when the pressure is very low it looks like a faint nebulous band over which brighter patches are superposed."

By a series of procedures which will not be detailed here, Thomson convinced himself "that the straight band of phosphores-

Fig. 2-5. An early low-pressure pattern. [*Phil. Mag.* **13** (1907), p. 573, Fig. 11.]

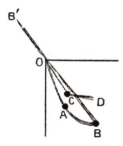

Fig. 2-6. Further development of the low-pressure pattern. [*Phil. Mag.* **20** (1910), p. 754, Fig. 3.]

cence which alone is seen at higher pressure and which lingers on with diminished intensity when the pressure is reduced, . . . is due [mostly] to secondary rays produced after the primary rays have passed through the cathode." He concentrated further attention on the phosphorescence due to the primary rays, and found that "when the pressure is low there seem with these large tubes to be parabolic bands corresponding to every gas in the tube." By measuring the parameters of these parabolas, and knowing the electric and magnetic fields used, he was able to compute from Eq. (2-9) the values of e/m, and he recognized the possibility that "this effect may furnish a valuable means of analysing the gases in the tube and determining their atomic weights."

However, he concluded that the method of recording patterns by transfer from the phosphorescent screen was unsatisfactory; in particular, "accurate measurements of the dimensions and positions of the luminous curves are more difficult than they would be on a photograph." He therefore shifted to photographic methods. ". . . I attempted to photograph the luminosity on the screen," but abandoned that method because of the long exposure times needed even with a very large lens. He found that "a much more sensitive and expeditious method is to put a photographic plate inside the tube itself and let the positive rays fall directly on the plate. . . . The photographic plate is very sensitive to these rays, . . . much more . . . than the . . . screen, and an exposure of three minutes shows curves on the plate which cannot be detected on the screen." He and a co-worker, F. W. Aston, devised ingenious methods of holding the plates, one of which is shown in Fig. 2-7. "In this method the plate is suspended by a silk thread wound round a tap which works in a ground-glass joint; by turning the tap the silk can be rolled or un-rolled and the plate lifted up or down. The plate slides in a vertical box of thin metal, light-tight except for the opening A, which comes at that part of the tube through which the positive rays pass. . . . When the silk is wound up, the strip of photographic plate in the box is above the opening, so that there is a free way for the positive rays to pass through the opening and fall on a . . . screen behind it, so that the state of the tube with respect to the production of positive rays can easily be ascertained. The box is large enough to hold a film long enough for three photographs."

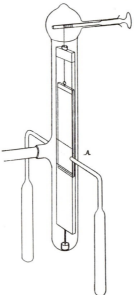

Fig. 2-7. One of the photographic plate holders. [*Phil. Mag.* **21** (1911), p. 228, Fig. 2.]

Thomson also arranged to keep the pressure in the deflection region much lower than that in the discharge by having the pump connected directly to the former, but to the latter only by way of the cathode channel. With these arrangements, a much larger variety of data was acquired; the seventh paper includes five photographs used for verifying atomic weights, as well as several others pertaining to various aspects of behavior of the rays. The method was still none too accurate. For example, the photograph obtained with carbon dioxide as the discharge gas showed one parabola corresponding to "an atomic weight about 47," but Thomson could not be sure whether this was actually due to CO_2 itself, which would have given 44, or to ozone, 48, formed in the discharge.

Thomson continued to improve the tubes themselves. The eighth paper of the series, for example, devotes considerable attention to studies of the shape of the cathode and of the discharge bulb in its vicinity.[4] The final version is shown in Fig. 2-8.

As has been mentioned, Thomson was interested in the possibility of using this procedure as a means of chemical analysis, and he seems to have looked for ways to prove its utility. Sir James Dewar

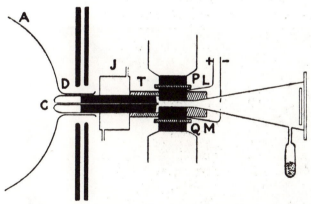

Fig. 2-8. Final form of the tube as used by Thomson. "A is a large bulb . . . in which the discharge passes, C the cathode placed in the neck [D] of the bulb. . . . The cathode is fixed into the glass vessel by a little wax. . . . The wax joint is surrounded by a water jacket J to prevent the wax being heated by the discharge. . . . An ebonite tube is turned to have the shape shown [at T (?)], L and M are two pieces of soft iron with carefully worked plane faces, placed so as to be parallel to each other, these are connected with a battery of storage cells and furnish the electric field. P and Q are the poles of an electromagnet separated from L and M by the thin walls of the ebonite box; when the electromagnet is in action there is a strong magnetic field between L and M." [*Proc. Roy. Soc., Ser. A* **89** (1913), p. 3, Fig. 4.]

"kindly supplied . . . two samples of gases obtained from the residues of liquid air"; and Thomson described the results of analyzing them before the Cambridge Philosophical Society and the Royal Institution. Of the two samples, one contained the heavier constituents and the other the lighter. The former, showing nothing exceptional, gave "an example of the convenience of the method, for a single photograph of the positive rays reveals at a glance the gases in the tube."

The other sample was more surprising. In "the photograph of the lighter constituents shown in Fig. [2-9] . . . we find the lines of helium, of neon (very strong), of argon, and in addition there is a line corresponding to an atomic weight 22, which cannot be identified with the line due to any known gas. I thought at first that this line, since its atomic weight is one-half that of CO_2, must be due to a carbonic acid molecule with a double charge of electricity, and on some of the plates a faint line at 44 could be detected. On passing the gas slowly through tubes immersed in liquid air the line at 44 completely disappeared, while the brightness of the one at 22 was not affected.

"The origin of this line presents many points of interest; there are no known gaseous compounds of any of the recognized elements

Fig. 2-9. The parabola spectrum of the lighter inert gases. [*Proc. Roy. Inst.* **20** (1913), facing p. 596, Fig. 3.]

which have this atomic weight. Again, if we accept Mendeleef's Periodic Law, there is no room for a new element with this atomic weight.... [I]t may possibly be a compound of neon and hydrogen, NeH_2, ... but ... I do not wish to lay much stress on this point.[5] There is, however, the possibility that we may be interpreting Mendeleef's law too rigidly, and that in the neighbourhood of the atomic weight of neon there may be a group of two or more elements with similar properties." In a lecture at the Royal Society a few months later, after having obtained the same sort of results from some very carefully purified specimens of neon, he was ready to put the matter somewhat differently: "There can, therefore, I think, be little doubt that what has been called neon is not a single gas but a mixture of two gases, one of which has an atomic weight about 20 and the other about 22."

For some reason, Thomson did not go so far as to suggest that there were two "kinds" of neon. Such an idea was definitely in the air at the time. The laws of radioactive decay were being enunciated by K. Fajans, A. S. Russell, and Frederick Soddy; in particular, Soddy had in February of 1913 presented before the Royal Society a paper which contained the following passage: "These results prove that almost every vacant place in the Periodic Table between thallium and uranium is crowded with non-separable[6] elements of atomic weight varying over several units, and leads inevitably to the presumption that the same may be true in other parts of the table. As previously pointed out, nothing further is necessary to explain the failure of all attempts to obtain numerical relations between the atomic weights. The view that the atomic mass is a real constant fixing all the chemical and physical properties of the elements is combatted most definitely by the fact that after the α-particle of mass 4 is expelled the members revert later to the original chemical type." It was Soddy, later in the same year, who proposed the name *isotopes*. Moreover, Soddy and others recognized that the radioactive transmutation laws would imply different atomic weights for lead of different radiogenic origins, and he and H. Hyman, as well as T. W. Richards and M. E. Lembert at Harvard, proceeded to verify this prediction.

If Thomson was aware of this work, he took no public notice of it.[7] Instead, he tried a different line of approach. If indeed the two lines were due to different "kinds" of neon, it ought to be possible to separate them by other physical processes depending on their mass (see note 6), and Aston followed this tactic. He first tried fractional distillation, with indefinite results. He next used diffusion, and on the first attempt got positive though not conclusive results.[8] A later run, however, seemed to give no effect.

There the matter was left until 1919.[9] Then Aston developed an electromagnetic separator which made true mass spectrometry possible. One of his first studies with the new instrument was the examination of neon, and this time there was no doubt: There were two (actually three[10]) varieties, or isotopes, of neon. Aston went on to demonstrate the presence of more than one isotope in most elements, and to determine highly accurate values for their masses. But the impetus had come from the extra parabola in Thomson's discharge tube.

Notes

1. This does not require the assumption that they are streams of distinct particles, but only that they are composed of some sort of ponderable matter, such as a charged fluid. Nevertheless, by 1907, at least for Thomson, the assumption that the rays have an atomic character was so automatic that it was never specifically mentioned.

2. This provides an example of the problems of style mentioned in a previous paragraph: The region that is referred to here as "behind the cathode" is the same as that described in the preceding sentence as "in front of the cathode."

3. The reader with a knowledge of calculus can see the origin of this as follows: In Eq. (2-2), $1/r$ is approximated by d^2y/dz^2, an approximation comparable to that of neglecting y'^2 in deriving Eq. (2-3). The flux density B is now a function of z, which can be written as $B_0 F(z)$. Therefore, Eq. (2-2) is replaced by

$$\frac{d^2y}{dz^2} = \frac{e}{mv} B_0 F(z).$$

Integration once with respect to z from $z = 0$ to $z = z'$ gives

$$\left.\frac{dy}{dz}\right|_{z'} - \left.\frac{dy}{dz}\right|_0 = \frac{e}{mv} B_0 \int_0^{z'} F(z)\,dz\,;$$

since the initial path is along the z axis, the second term on the left-hand side is zero. The resulting equation can be integrated a second time, with respect to z', from 0 to z; using the fact that $y = 0$ for $z = 0$ gives

$$y = \frac{eB_0}{mv} \int_0^z dz' \int_0^{z'} F(z)\,dz.$$

Thus $f(z) = \int_0^z dz' \int_0^{z'} F(z)\,dz$. It will also be recognized that for $F(z) = 1$, which represents the case of a uniform field, this gives just Eq. (2-4).

4. This paper also included an account of some work done with an electric detection scheme, which measured the charge delivered by rays of a

chosen value of e/m. Comparison of this with methods using either phosphorescence or photography showed that these methods were not useful for judging relative amounts of ions of different kinds, as they responded to something approximating the velocity of the incident particles as well as to their number. In particular, it turned out that lines due to hydrogen, which were practically unavoidable on the screen or plate, represented only tiny amounts of hydrogen relative to the amounts of primary ions. (It was later recognized that hydrogen and helium were driven into the parts of the tube when used as the discharge gas, and could be completely removed only by a baking process that is impossible with glass systems sealed with wax and grease.)

5. In addition to the chemical inertness of neon, Thomson was influenced by the fact that the new substance also appeared frequently with a mass-to-charge ratio of 11, i.e., with a double charge. Such an occurrence was rare for known molecules but common for atoms.

6. Soddy's use of this term led to an amusing exchange of letters between Soddy and Professor Arthur Schuster, of Manchester, in the pages of *Nature*. Soddy had to concede that the difference in masses would permit separation by physical, though not by chemical, processes. The interchange is given in *World of the Atom*, vol. I, pp. 784, 786.

7. He was relatively uninterested in radioactivity; he is on record as feeling that radioactive atoms were so special that very little of what was learned from them could be applied to ordinary atoms.

8. These results were reported at a meeting of the British Association in September 1913, and attracted a large audience. The session was apparently concurrent with one at which Soddy was expounding his laws of radioactive transmutation! A report in *Nature* of Soddy's presentation is given in *World of the Atom*, vol. I, p. 790. It is remarkable that Soddy seems to have taken as little note of Thomson's work at this time as Thomson did of his.

9. World War II was the major reason for this lapse.

10. The third, of mass 21, was too rare to show up in Thomson's work.

Bibliography

Basic papers: Sir J. J. Thomson, *Philosophical Magazine* **13**, 561 (1907); **14**, 359 (1907); **16**, 657 (1908); **18**, 821 (1909); **19**, 424 (1910); **20**, 752 (1910); **21**, 225 (1911); and **24**, 209 (1912); *Proceedings of the Cambridge Philosophical Society* **17**, 201 (1914). Frederick Soddy, *Chemical News* **107**, 97 (1913); *Nature* **92**, 399 (1913). F. W. Aston, *Philosophical Magazine* **39**, 449 (1920).

Summaries: Sir J. J. Thomson, *Proceedings of the Royal Institution* **20**, 140, 591 (1911-1913); *Proceedings of the Royal Society*, *Ser. A* **89**, 1 (1913); *Rays of Positive Electricity*, 2nd ed. (Longmans, Green and Company, London, 1921). An abridgement of the paper in the *Proceedings of the Royal Society* is given in *World of the Atom*, vol. I, p. 795.

See also F. W. Aston, *Isotopes* (Pergamon Press, New York, 1955).

Chapter 3

The Meaning of
Atomic Number

By the latter part of 1913, there was no longer any serious opposition to the atomic hypothesis, and attention had turned, rather, to the problem of deducing the structure of the atom. Here significant progress had been made, with strong evidence favoring Rutherford's nuclear model.[1] However, a basic parameter in any attempt (such as that of Bohr) to elaborate that model was the charge carried by the nucleus, and here there was considerable uncertainty. Experiments on the scattering of x rays, carried out in 1907 by Charles Barkla, had established that the nuclear charge was roughly half the atomic weight times the magnitude of the charge on the electron; Geiger and Marsden's work on the scattering of α particles had shown agreement with that approximation; and Bohr had used it with striking success. But it was still only an approximate result, reliable to only about 5%. The resolution of the uncertainty came, not from an attack aimed directly at it, but rather from a systematic study of x rays by Henry G. J. Moseley. In two papers in the *Philosophical Magazine* in 1913 and 1914, he showed that the proper measure of the nuclear charge is the atomic number, the number giving the order of the element in the Mendeleev periodic system.

Barkla and C. A. Sadler had originally recognized that each element, if used as the emitting target in an x-ray tube, gave off a spectrum composed of two parts: a continuum, which did not depend on the nature of the target, and a set of discrete "lines,"[2] which were as distinctively characteristic of the element as were the lines in its optical spectrum. They found that the characteristic radiations from any given element formed groups, which were labeled K, L, . . . , in order of decreasing hardness, with the hardness of each group increasing from element to element as one went higher and higher in the periodic table. The methods used by Barkla and Sadler did not permit them to obtain any more definite relationships. Accordingly, shortly after the wave nature of x rays had been established and the measurement of their wavelengths had become possible, Moseley, starting at the University of Manchester and continuing at Oxford, undertook "to make a general survey of the principal types of high-frequency radiation."

Moseley made use of a method of analysis developed by W. H. Bragg and W. L. Bragg on the basis of the younger Bragg's analysis of the work of Laue, Friedrich, and Knipping (Chap. 2): the reflection of a beam of x rays from a cleavage plane of a crystal. In this arrangement, as Moseley expressed it, "The radiations of definite frequency . . . are reflected only when they strike the surface at definite angles, the glancing angle of incidence θ, the wavelength λ, and the 'grating constant' d of the crystal being connected by the relation

$$n\lambda = 2d \sin \theta,$$

where n, an integer, may be called the 'order' in which the reflexion occurs." A plan of the basic features of the arrangement is shown in Fig. 3-1. "The X rays, after passing through the slit marked S, . . . fell on the cleavage face, C, of a crystal of potassium ferrocyanide which was mounted on the prism-table of a spectrometer. The surface of the crystal was vertical and contained the geometrical axis of the spectrometer."

Previous applications of the method, by the Braggs and by Moseley and C. G. Darwin, had involved electrical detection methods based on the ionization produced by the x rays. Moseley noted that this method is "only successful where a constant source of radiation is available," and he therefore devised a means of using photographic detection.

"The photographic plate was mounted on the spectrometer arm, and both the plate and the slit were 17 cm. from the axis. The importance of this arrangement lies in a geometrical property, for when these two distances are equal the point L at which a beam reflected at a definite angle strikes the plate is independent of the position of [the point of incidence] P on the crystal. The angle at which the crystal is set is then immaterial so long as a ray can strike some part

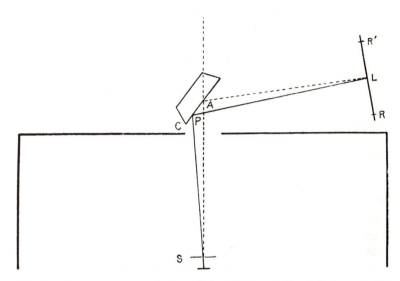

Fig. 3-1. Plan of the arrangement of crystal, source slit, and photographic plate in Moseley's apparatus. [*Phil. Mag.* **26** (1913), p. 1025, Fig. 1.]

of the surface at the required angle. The angle θ can be obtained from the relation $2\theta = 180° - SPL = 180° - SAL$.

"The following method was used for measuring the angle SAL. Before taking a photograph a reference line R was made at both ends of the plate by replacing the crystal by a lead screen furnished with a fine slit which coincided with the axis of the spectrometer. A few seconds' exposure to the x rays then gave a line R on the plate, and so defined on it the line joining S and A. A second line R' was made in the same way after turning the spectrometer arm through a definite angle. The arm was then turned to the position required to catch the reflected beam and the angles LAP for any lines which were subsequently found on the plate deduced from the known value of RAP [the amount by which the arm had been rotated to catch the reflected beam] and the position of the lines on the plate. The angle LAR was measured with an error of not more than $0°.1$, by superimposing on the negative a plate on which reference lines had been marked in the same way at intervals of $1°$. In finding from this the glancing angle of reflexion two small corrections were necessary in practice, since neither the face of the crystal nor the lead slit coincided accurately with the axis of the spectrometer."

The order n "was determined by photographing every spectrum both in the second order and in the third"; this also provided a check on the accuracy. The crystal structure of potassium ferrocyanide was not known at the time, and therefore the grating constant d could not be calculated; but the particular crystal had been calibrated against a crystal of rocksalt. W. L. Bragg had shown that the atoms in rocksalt form a simple cubical array, so that its grating constant could be computed from its density and molecular weight and the natural constants N_0, Avogadro's number, and e, the charge on the electron. The values of the latter two, especially e, were not too reliably known, but the value of d depends on the cube roots of both so that the errors were not serious.[3]

The arrangement for interchanging targets is shown in Figs. 3-2 and 3-3. "The aluminium trolley which carries the targets can be drawn to and fro by means of silk fishing-line wound on brass bobbins. [The bobbins were rotated by means of the ground glass joints on the vertical side tubes at each end.] An iron screen S fastened to the rails is furnished with a fine vertical slit which defines the X-ray beam. The slit should be fixed exactly opposite the focus-spot of the cathode-stream [the x rays were excited by cathode-ray bombardment of the targets], though a slight error can be remedied by deflecting the cathode rays with a magnet. The X rays escape by a side-tube 2½ cm. diameter closed by an aluminium window 0.022 mm. thick." The spectrometer was shielded from the rest of the radiation by a lead box surrounding the tube, as indicated in Fig. 3-1.

For the longer wavelengths dealt with in the second paper, there were additional complications, as these radiations "cannot penetrate an aluminium window or more than a centimetre or two of air. The photographs had therefore to be taken inside an exhausted spectrometer." Fig. 3-3 pertains to this arrangement. The x-ray tube "consists of a bulb containing the cathode, joined by a very large glass T-piece to a long tube of 4 cm. diameter, in which are the rails R and the carriage C. S is the defining-slit and W a window of gold-beaters' skin[4] which separates the tube from the spectrometer. . . . The spectrometer, which was specially designed for this work, consists of a strong circular iron box of 30 cm. inside diameter and 8 cm. high, closed by a lid which, when the flange is greased, makes an air-tight joint. Two concentric grooves are cut in the floor of the box. The table A, which carries the plate-holder, rests on three steel balls, of which two run in the outer groove, while the third rests on the floor of the box. The position of the crystal-table is controlled in like manner by the inner groove. . . . The scales are fixed to the box and the verniers[5] to the tables."

The use of the vacuum box produced other problems: "In this case the slit and photograph are not equidistant from the crystal, and the position of the spectrum-lines on the plate is no longer independent of the angle at which the crystal is set. The necessary corrections were calculated geometrically, and verified by photographing the

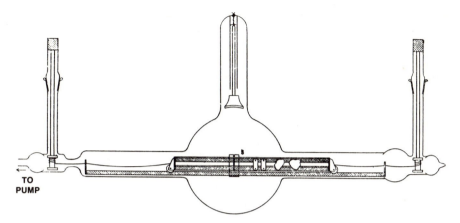

Fig. 3-2. Drawing of Moseley's x-ray tube, showing the arrangement for interchanging targets. [*Phil. Mag.* **27** (1914), p. 704, Fig. 1.]

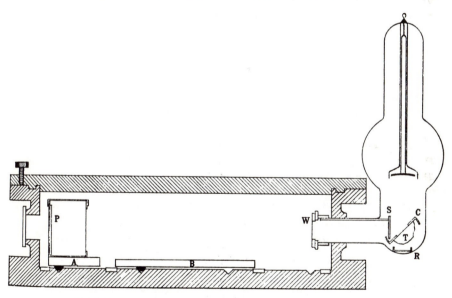

Fig. 3-3. The x-ray tube in end section, showing the rails, carriage, target, slit screen, and vacuum spectrometer box. [*Phil. Mag.* **27** (1914), p. 705, Fig. 2.]

same line for both right-handed and left-handed reflexions and with the crystal set at various angles."

Moseley states, "The only serious difficulty in the experiments is caused by the heat produced by the cathode ray bombardment, and the consequent liberation of gas and destruction of the surface of the target. This makes it necessary to use the element in a form which is not too volatile and prevents the employment of a very powerful discharge. The total time of an exposure, including rests, varied from three minutes for a substance such as ruthenium, which could safely be heated, to thirty minutes for the rare earth oxides."

There was also the problem of interference from the continuous radiation. For most of the work, this caused no trouble, as its spectrum was such that it was reflected at angles much smaller than the lines of interest. However, "in the work on the very short wavelengths . . . [such as the spectrum] of Ag in the K series, the general reflexion cannot be avoided. Unfortunately, when photographed it takes the form of irregular fringes, which effectually hide faint spectrum-lines." The fringes were known to be due to a "pattern of patches on the crystal surface which reflect exceptionally well," and this fact made it "easy to devise methods for getting rid of the fringes. In the first place, narrowing the slit or increasing the distance from the crystal will diminish their intensity compared with that of the line-spectrum. In the second place, turning the crystal will move and blur the fringes, but leave the sharpness of the lines unaffected provided the slit and photograph are equidistant from the reflecting surface"—a condition which, it will be recalled, was met in the work with short wavelengths.

In all, spectra were recorded for 38 elements ranging from aluminum to gold. Of these, 15 (Al through Y) were studied in the K series only, 17 (Sn through Au) in the L series only, and 6 (Zr, Nb, Mo, Ru, Pd, and Ag) in both. Table 3-I gives the results for the K series, and Fig. 3-4 shows the spectra of ten of these elements placed "approximately in register," so that "those parts of the photographs which represent the same angle of reflexion are in the same vertical line." The regularity of the progression in wavelength is quite obvious. Its shape apparently suggested to Moseley a quadratic curve, as he proceeded in the first paper to calculate the quantity given in the third column of Table 3-I,

$$Q_K = \sqrt{\nu / \tfrac{3}{4} \nu_0},$$

where ν is the frequency of the longer-wavelength line in each spectrum and ν_0 is "the fundamental frequency of ordinary line spectra . . . obtained from Rydberg's wave-number . . . = 109,720 [cm^{-1}]. . . . It is at once evident that Q increases by a constant amount as we pass from one element to the next, using the chemical order of the elements in the periodic system. . . . We have here a proof that there is in the atom a fundamental quantity, which increases by regular steps as we pass from one element to the next. This quantity can only be the charge on the central positive nucleus,

Table 3-I. Moseley's results for the K series x rays. [*Phil. Mag.* **27** (1914), p. 708, Table I.]

	α line λ × 10⁸ cm	Q_K	N Atomic number	β line λ × 10⁸
Aluminum	8.364	12.05	**13**	7.912
Silicon	7.142	13.04	14	6.729
Chlorine	4.750	16.00	17	. . . . .
Potassium	3.759	17.98	19	3.463
Calcium	3.368	19.00	20	3.094
Titanium	2.758	20.99	22	2.524
Vanadium	2.519	21.96	23	2.297
Chromium	2.301	22.98	24	2.093
Manganese	2.111	23.99	25	1.818
Iron	1.946	24.99	26	1.765
Cobalt	1.798	26.00	27	1.629
Nickel	1.662	27.04	28	1.506
Copper	1.549	28.01	29	1.402
Zinc	1.445	29.01	30	1.306
Yttrium	0.838	38.1	39	. . . . .
Zirconium	0.794	39.1	40	. . . . .
Niobium	0.750	40.2	41	. . . . .
Molybdenum	0.721	41.2	42	. . . . .
Ruthenium	0.638	43.6	44	. . . . .
Palladium	0.584	45.6	46	. . . . .
Silver	0.560	46.6	47	. . . . .

of the existence of which we already have definite proof. Rutherford has shown, from the magnitude of the scattering of α particles by matter, that this nucleus carries a + charge approximately equal to A/2 electrons, where A is the atomic weight. Barkla, from the scattering of X rays by matter, has shown that the number of electrons in an atom is roughly A/2, which for an electrically neutral atom comes to the same thing. Now atomic weights increase on the average by about 2 units at a time, and this strongly suggests that N increases from atom to atom always by a single electronic unit. We are therefore led by experiment to the view that N is the same as the number of the place occupied by the element in the periodic system."

The extension of Moseley's work in the second paper made the case even stronger. "In Fig. [3-5] the spectra of the elements[6] are arranged on horizontal lines spaced at equal distances. The order chosen for the elements is the order of the atomic weights, except in the cases of A, Co, and Te, where this clashes with the order of the chemical properties.[7] Vacant lines have been left for an element between Mo and Ru, an element between Nd and Sa, and an element

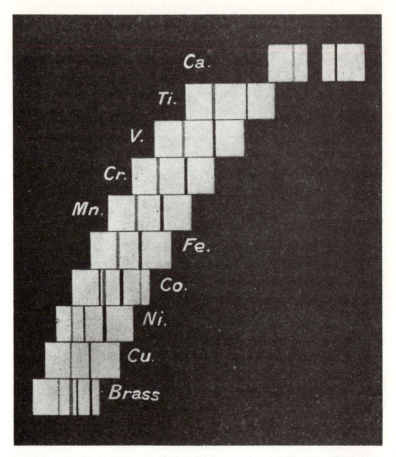

Fig. 3-4. The K x-ray spectra of ten elements between Ca and Zn. [*Phil. Mag.* **26** (1913), Plate XXIII.]

between W and Os, none of which are yet known, while Tm, which Welsbach has separated into two constituents, is given two lines.[8] This is equivalent to assigning to successive elements a series of successive positive integers. On this principle the integer N for Al, the thirteenth element, has been taken to be 13, and the values of N then assumed by the other elements are given on the left-hand side of fig. [3-5]. This proceeding is justified by the fact that it introduces perfect regularity into the X-ray spectra. Examination of fig. [3-5] shows that the values of $\nu^{1/2}$ for all the lines examined both in the K and L series now fall on regular curves which approximate to straight lines. . . .

"Now if either the elements were not characterized by these integers, or if any mistake had been made in the order chosen or in

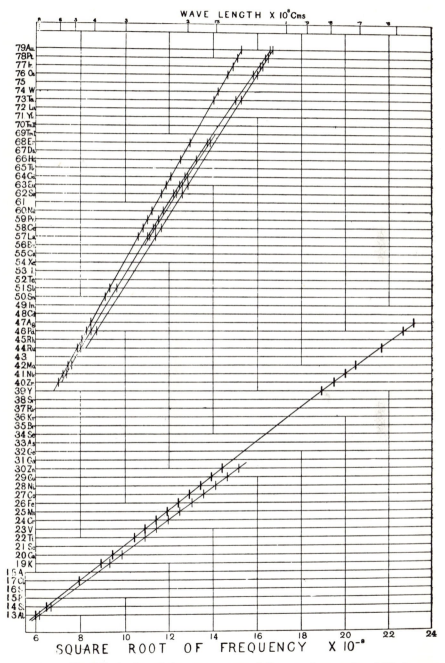

Fig. 3-5. A plot of the square roots of the x-ray frequencies of 38 elements against their positions in the periodic system. [*Phil. Mag.* **27** (1914), p. 709, Fig. 3.]

the number of places left for unknown elements, these regularities would at once disappear.[9] We can therefore conclude from the evidence of the X-ray spectra alone, without using any theory of atomic structure, that these integers are really characteristic of the elements. . . .

"Now Rutherford has proved that the most important constituent of an atom is its central positively charged nucleus, and van den Broek has put forward the view that the charge carried by this nucleus is in all cases an integral multiple of the charge on the hydrogen nucleus. There is every reason to suppose that the integer which controls the X-ray spectrum is the same as the number of electrical units in the nucleus, and these experiments therefore give the strongest possible support to the hypothesis of van den Broek. Soddy has pointed out that the chemical properties of the radioelements are strong evidence that this hypothesis is true for the elements from thallium to uranium, so that its general validity would now seem to be established."

Notes

1. See *Crucial Experiments*, chap. 5.

2. The term is used here despite the fact that at the time, the spectrum was a distribution in "hardness," that is, resistance to absorption, and not in wavelength, and the pattern of intensity was represented by heights of a curve rather than lines on a photograph.

3. As a matter of fact, they would cancel out of the relative values anyway; they would cause all absolute wavelength (or frequency) values to be in error by a single factor.

4. This refers to an animal membrane used to separate the metal foils in making gold leaf. It is "usually air-tight, though sometimes it may require varnishing," and is "extremely transparent to X rays."

5. This refers to Vernier scales for measuring angles.

6. The K spectra had already been found to consist of two lines; the L spectra consisted of up to six lines, of which four were systematically measured.

7. Moseley had already noted in the first paper that "A is itself probably a complicated function of N," and that "atomic weights vary in an apparently arbitrary manner, so that an exception in their order does not come as a surprise."

8. All three of the missing elements have since been discovered; only one of them, Re, which comes between W and Os, occurs naturally. The chemistry of the rare earth elements is notoriously difficult; there is no reason to doubt Welsbach's "separation" of Tm, but also no reason to ascribe the two components adjacent places in the periodic system. Some other assignments in this region have been revealed by later developments to be erroneous.

9. This is slightly exaggerated. It is now known that the atomic number of Ho, for which wavelength data are included in the figure, is 67 rather than 66.

Bibliography

Moseley's papers appeared in *Philosophical Magazine* **26**, 1024 (1913), and **27**, 703 (1914); they are excerpted in *World of the Atom*, vol. II, p. 874.

The original suggestion by Van den Broek, mentioned on p. 36, was made in a Letter in *Nature* **92**, 372 (1913), which is reprinted in *World of the Atom*, vol. I, p. 856.

Chapter 4

Superconductivity

The liquefaction of helium was first accomplished in 1908 by Heike Kamerlingh Onnes of the University of Leiden, and for the first time it became possible to study phenomena within a few degrees of absolute zero (the boiling point of helium at atmospheric pressure is 4.2 K). One line of research concerned the variation of the resistivity of metals with temperature. Kamerlingh Onnes had already carried out such studies down to the temperature of liquid air, about 80 K. He had found for several pure metals an approximately linear relationship which he realized could not continue indefinitely, as it would imply negative resistance at absolute zero. Sir James Dewar had continued the work, reaching liquid hydrogen temperatures (20 K), and had found that indeed the variation became less rapid.

This was as expected, not only for the reason already adduced, but also on the basis of the ideas then held about metals and their properties. Electrical conduction was supposed to be mediated by electrons, with resistance arising from collisions between the electrons and the atoms of the metal. The linear decrease was in accordance with what was believed about the variation of the electron movements with temperature. At low enough temperatures, however, the electrons were expected to "condense" on the atoms, leading to a minimum in the resistivity and an eventual conversion of the metal into an insulator.

The actual behavior was drastically different. Kamerlingh Onnes found that for many metals, the resistivity approached a constant value as the temperature was lowered; and for some, the resistance disappeared entirely at a characteristic temperature which, it turned out, depended on the magnetic field. These experiments were among those for which Kamerlingh Onnes was awarded the Nobel prize for physics in 1913.

For over twenty years, the vanishing of resistivity was regarded as the essential feature of superconductivity. However, this had some puzzling features. If a magnetic field is applied to an ordinary conductor (other than a ferromagnetic material), some of the flux will pass through the conductor; on the other hand, if a magnetic field is applied to a perfect conductor, surface currents will be induced such as to produce a magnetic intensity inside the conductor that just cancels the applied field and keeps the flux inside the conductor zero. This means that the state of a superconductor in a magnetic field depends on the history of the situation,[1] a highly unsatisfactory state of affairs. Finally, in 1933, Walther Meissner, R. Ochsenfeld, and F. Heidenreich showed that a metal that becomes a superconductor actually expels the flux if the temperature is lowered through the transition value while the specimen is in a magnetic field. Both the work of Kamerlingh Onnes and that of Meissner, Ochsenfeld, and Heidenreich are described in this chapter.

Kamerlingh Onnes's liquefier is shown in Fig. 4-1, which consists of a photograph in part (a) and a schematic drawing in part (b). Much of the lettering refers to details of operation of the device which are of no concern here. The essentials are contained in the flow path of the helium, which passes in the following sequence:

"(a) through a tube Ca which at its lower end is cooled down far below the freezing point by means of vapor of liquid air, and at its upper end is kept at the ordinary temperature. Here the helium is perfectly dried.

"(b) through a tube divided into two parts along two refrigerating tubes (in Da and Db), in which it is cooled in one by the abduced[2] hydrogen, in the other by the abduced helium, after which it unites again.

"(c) through a tube Cb filled with exhausted charcoal and immerged in liquid air. Here whatever traces of air might have been absorbed during the circulation, remain behind.

"(d) through a refrigerating tube B_3 lying in the liquid air which keeps the cover of the hydrogen space and of the helium space cooled down.

"(e) through a refrigerating tube B_2, in which it is cooled by the evaporated liquid hydrogen.

"(f) through the refrigerating tube B_1 lying in the liquid hydrogen evaporating under a pressure of 6 cm., here the compressed helium is cooled down to $15°$K;

"(g) and from here in the regenerator coil A. . . .

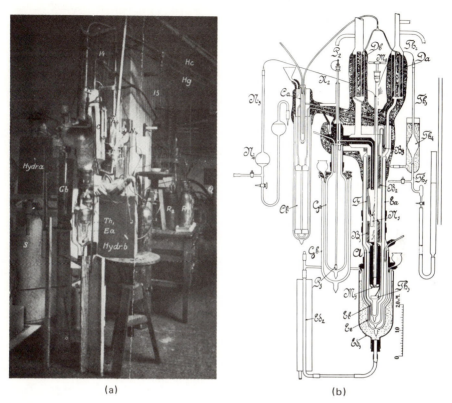

(a)   (b)

Fig. 4-1. (a) Photograph of Kamerlingh Onnes's helium liquefier. [*Commun. Phys. Lab. Univ. Leiden* No. 108 (1908), Plate I.] (b) Schematic diagram of the heart of the liquefier. The letters in both parts of the figure are explained in the text. [*Commun. Phys. Lab. Univ. Leiden* No. 108 (1908), Plate III.]

"Then it expands through the cock M_1."

This last is the key step in the process: When a gas is allowed to expand through an orifice from a region of high pressure to one of lower pressure, its temperature decreases, provided it is initially already below a critical value, the *inversion temperature*, that depends on the gas.[3] To obtain liquefaction of any significant fraction of the gas, it must initially be below about one-third of the inversion temperature. For helium, the inversion temperature is 51 K; it is this fact that makes necessary all the preliminary cooling stages, and explains why the very achievement of liquefaction was a major feat.

"... When the temperature has descended so low that the liquid helium flows out, the latter collects in the lower part of the vacuum glass *Ea*. ... Round the transparent bottom part of the vacuum glass a protection of liquid hydrogen has been applied. The second vacuum glass Eb, which serves this purpose, forms a closed space together with the former Ea, and the construction has been arranged in such a way that first this space can be exhausted and filled with pure hydrogen gas, which is necessary to keep the liquid hydrogen perfectly clear later on.... The hydrogen glass is surrounded by a vacuum glass Ec with liquid air, which in its turn is surrounded by a glass Ed with alcohol, heated[4] by circulation.

"By these contrivances and the extreme purity of the helium we succeeded in keeping the apparatus perfectly transparent to the end of the experiment, after 5 hours."

The first resistance measurements were made in a modification of the original liquefier, diagrammed in Fig. 4-2. The specimen was a platinum-wire resistance that had been used in previous work and is shown as "Ω" in the figure. "The thin platinum wire is wound round a glass cylinder and is kept tight on it by being wound while hot, and the thicker platinum ends W_a and W_b are fused to the glass. To these ends the double platinum leads W_{a1}, W_{a2} and W_{b1}, W_{b2} are ... tin-soldered." These leads are also labeled where they leave the apparatus at the upper right. "The resistance was measured on the Wheatstone bridge.... [I]t appears that by descending to helium temperatures the resistance is still further diminished, but when these temperatures are reached the resistance attains a constant value quite independent of the individual temperature to which it has been brought. The results are plotted in [Fig. 4-3], which shows well the asymptotical approach of the resistance to a constant value at $4.3°\,$K." Also shown are earlier results from two specimens of gold of different and known purities. The comparison of the behavior of these curves led Kamerlingh Onnes to believe that for a truly pure metal, the resistivity would vanish above the boiling point of helium in the manner indicated by the lowest curve in Fig. 4-3.

The next step was the construction of a vessel, called a *cryostat*, into which the liquid helium could be siphoned and which could accommodate a stirrer (to ensure uniformity of temperature) as well as larger experimental specimens. The first resistance measurements made with this arrangement were on gold and mercury. For the latter, the results showed a value at 3 K of less than 0.0001 times that at $0°C,$[5] and for the former, they indicated that the resistance did indeed vanish; but they were admittedly imprecise. Since it was believed that at least part of the residual effects were due to impurities and since mercury could be purified much more easily than gold, further attention was concentrated on mercury. The next measurements, more precise, not only confirmed the previous conclusions, but "it has also been ascertained that the actual value of the

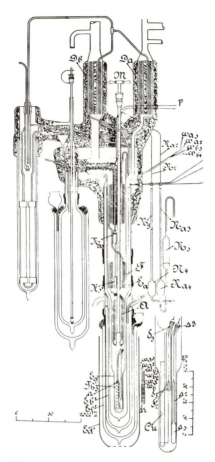

Fig. 4-2. Modification of the liquefier for resistance measurements. The letters are the same as in Fig. 4-1. The lines leading off to the right represent connections to auxiliary apparatus used in another experiment. [*Commun. Phys. Lab. Univ. Leiden* No. 119 (1911), Plate I, Fig. 1, with a portion removed.]

resistance is very much smaller than [the] upper limit which I was able to ascribe to it from my former measurements."

A typical mercury specimen and its arrangement in the cryostat are shown in Fig. 4-4. "Seven glass U-tubes of about 0.005 sq. mm. cross section are joined together at their upper ends by inverted Y-pieces which are sealed off above, and are not quite filled with mercury; this gives the mercury an opportunity to contract or expand on freezing or liquefying without breaking the glass and without breaking the continuity of the mercury thread formed in the seven U-tubes. To the Y-pieces b_0 and b_8 are attached two leading tubes

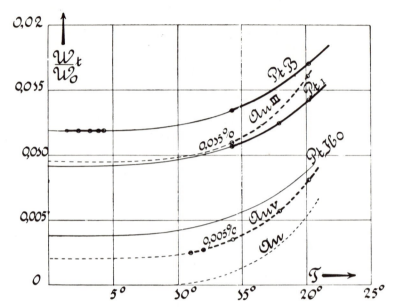

Fig. 4-3. Resistances of two platinum specimens and two gold specimens; the thin portions of the latter are extrapolations. [*Commun. Phys. Lab. Univ. Leiden* No. 119 (1911), Plate III, Fig. 3.]

Hg_1, Hg_2 and Hg_3, Hg_4 (whose lower portions are shown at Hg_{10}, Hg_{20}, Hg_{30}, Hg_{40}) filled with mercury which, on solidification, forms four leads of solid mercury. To the connector b_4 is attached a single tube Hg_5, whose lower part is shown at Hg_{50}. At b_0 and b_8 current enters and leaves through the tubes Hg_1 and Hg_4; the tubes Hg_2 and Hg_3 can be used for the same purpose or also for determining the potential difference between the ends of the mercury thread. The mercury filled tube Hg_5 can be used for measuring the potential at the point b_4. To take up less space in the cryostat and to find room alongside the stirring pump Sb, the tubes which are shown in one plane in [part (a) of Fig. 4-4] were closed together in the manner shown in [part (b)]. The position in the cryostat is shown in [part (c)]. . . . The leads project above the cover Sb_1 in a manner shown in perspective in [part (d)]. They too are provided with expansion spaces, while in the bent side pieces are fused platinum wires $Hg_{1'}$, $Hg_{2'}$, $Hg_{3'}$, $Hg_{4'}$, $Hg_{5'}$ which are connected to the measuring apparatus. The apparatus was filled with mercury distilled over in vacuo at a temperature of $60°$ to $70°C$. while the cold portion of the distilling apparatus was immersed in liquid air.''

The results obtained were of the form shown in Fig. 4-5. "As a former experiment showed that there was a pretty rapid diminution

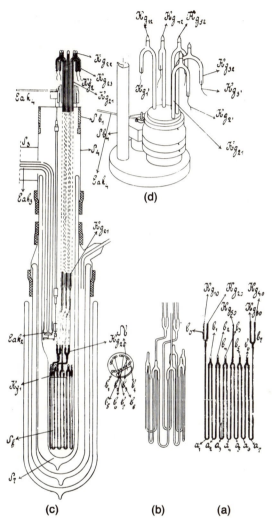

Fig. 4-4. (a) Schematic arrangement of a mercury specimen for resistance measurements. (b) Perspective drawing of the specimen. (c) Sketch of the cryostat with the specimen in position. (d) Cover of the cryostat, showing how the leads were brought out. [*Commun. Phys. Lab. Univ. Leiden* No. 124c (1912), Plate I, Figs. 1, 2, 4, and 3, respectively.]

of the resistance just below the boiling point of helium, there arose in the first place a question as to whether there exists . . . a point of inflection in the curve. . . . The temperature of the bath was therefore raised above the boiling point by allowing the pressure under which the liquid evaporated to increase . . . by closing the tap Eak_2,

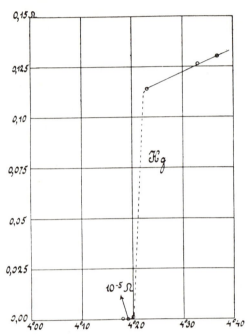

Fig. 4-5. Resistance of a mercury specimen as a function of the temperature. [*Commun. Phys. Lab. Univ. Leiden* No. 124 (1912), p. 23, unnumbered figure.]

leading to the liquefier. . . . These measurements showed that from the melting point of hydrogen to the neighbourhood of the boiling point of helium the curve exhibited the ordinary gradual lessening of the rate of diminution of resistance. . . . A little above and a little below the boiling point, from 4°.29 K. to 4°.21 K. the same gradual change was clearly evident (cf. the fig.), but between 4°.21 K. and 4°.19 K. the resistance diminished very rapidly and disappeared at 4°.19 K. (Temperature measurements are here given with 4°.25 K. as the boiling point of helium.)" The superconducting state was clearly established.

The next step was to study the new state with higher currents; the basis for this was that if the resistance was not truly zero, a larger current would give rise to a larger, and thus more readily detectable,[5] potential difference. The results only served to confuse the situation, as there appeared "special phenomena . . . that at every temperature below 4°.18 K. for a mercury thread inclosed in a glass capillary tube a '*threshold value*' of the current density can be given, such that at the crossing of the 'threshold value' the phenomena change. At current density below the 'threshold value' the electricity goes through without any perceptible [*sic*] potential difference at the ends of the thread being necessary. It appears therefore that the thread has no resistance. . . . As soon as the current density rises above the 'threshold value', a potential difference appears which *increases more rapidly than the current.*"

There followed a series of experiments designed to try to find the explanation for this new effect. It was first noted that the threshold current density was larger the lower the temperature, increasing approximately in proportion to the difference from the transition temperature as long as that difference was not very large. The natural assumption was that some sort of heating effect was taking place, raising the temperature of the mercury above the transition point, and the aim was to find out how the heat was produced. By various special configurations of threads, it was established that heat was not being conducted in from the outside. The effect of impurities in the mercury was considered, even though the distillation process by which it was prepared should have eliminated them; experiments showed that the effect was not caused by reasonable amounts of impurities. It was thought that perhaps contact with an ordinary conductor, somehow present or formed in the mercury thread, might nullify the superconducting property; this was tested by use of a steel capillary, with uncertain results, but later ruled out by a comparable experiment on tin. In sum, the work with mercury failed to provide the answer.

However, as Kamerlingh Onnes recognized, mercury was not a wholly satisfactory substance to work with. "Various circumstances combined to make . . . the investigation of mercury enclosed in capillary tubes difficult. A day of experiments with liquid helium requires a great deal of preparation, and when the experiments treated of here were made, . . . there were only a few hours available for the actual experiments. To be able to make accurate measurements with the liquid helium then, it is necessary to draw a programme beforehand and to follow it quickly and methodically on the day of experiment. Modifications of the experiments in connection with what one observes, must usually be postponed to another day. . . . Very likely in consequence of some delay caused by the careful and difficult preparation of the resistances, the helium apparatus would have been taken into use for something else. And when we could go on with the experiment again, the resistance sometimes became useless . . . because in the freezing the fine mercury thread separated, and all our preparations were labour thrown away. Under these circumstances the detection and elimination of the causes of unexpected and misleading disturbances took up a great deal of time."

In addition, it was desirable to cool the specimen by direct contact with the liquid helium rather than through the wall of a capillary tube. Consequently, "when I found that . . . tin and lead show similar properties to mercury, the investigations were continued with these two metals." It was then that the problem was solved. In fact, there was already a hint of the solution in the experiments in which lead was established as a superconductor. Lead could be formed into wires easily; a fairly considerable amount of wire $1/70$ mm^2 in cross-sectional area was made. With a single strand of this wire, the threshold current at 4.25 K was 8 amperes. Then a coil was wound with it, 1000 turns in layers 1 cm wide along a core 1 cm in diameter. The winding was insulated with silk, which became soaked in liquid helium. The threshold current for the coil was found to be only 0.8 ampere.

Already in 1913 there was widespread interest in the production of intense magnetic fields, and it had been recognized that a major problem was the power dissipated in the windings. Perrin, for example, had suggested the use of liquid air as a coolant, with the expected advantage of less heat produced because the low temperature would reduce the resistance of the windings; but calculations had shown this to be unfeasible, largely because of the problem of achieving adequate heat transfer between a supposedly compact coil and the coolant. Kamerlingh Onnes grasped the potentialities of superconductors for such an application, noting that they would not produce heat. However, he admitted in making this suggestion "the possibility that a resistance is developed in the superconductor by the magnetic field," and proceeded to investigate the question.

"There were . . . reasons to suppose that its amount would be small. . . . A direct proof that in supraconductors only an insignificant resistance was originated by the magnetic field was found in the fact that [the coil described above] remained supra-conducting, even when a current of 0.8 ampère was sent through it. The field of the coil itself amounted in that case to several hundred gauss, and a great part of the turns were in a field of this order of magnitude, without any resistance being observed." Kamerlingh Onnes therefore "arranged the apparatus for these experiments as if for a phenomenon that could only be studied with profit in fields of kilo-gauss." The results were again surprising.

A lead coil, which had been used in earlier studies and was known to become superconducting, was placed in the cryostat with the plane of the windings parallel to the magnetic field.

"It was first ascertained that the coil was supra-conducting at the boiling-point of helium. Further that it remained supra-conducting when a current of 0.4 ampère was sent through it; even then the windings were in a not inconsiderable field of their own current. . . .

"Then the magnetic field was applied. With a field of 10 Kilo-gauss there was a considerable resistance, at 5 Kilo-gauss it was somewhat less. This made it fairly certain that the magnetic field created resistance in supra-conductors at larger intensities, and not at smaller ones. . . . Further investigation gave for the resistance (expressed in parts of the resistance at $0°C$) as function of the field, the curve that is given diagrammatically completed in [Fig. 4-6]."

Kamerlingh Onnes was not yet prepared to relate the critical current to the critical magnetic field. (It is now known that for samples with dimensions larger than about 10^{-4} cm, the critical current is that current which produces a magnetic field at the surface equal to the critical field.) Also, while he had "no doubt that the phenomenon discovered here is connected with the sudden appearance of ordinary resistance in the supra-conductors at a certain temperature," it was left to others to establish the form of the relation.[6] Nevertheless, the foundations were well laid.

As time went on, however, the puzzle mentioned in the introductory paragraphs of this chapter became apparent. In fact, a slightly modified version makes the puzzle even more complex. If a material were to undergo a transition to a perfectly conducting state by reduction of temperature while in a magnetic field, the flux passing through it at the instant of the transition would be "frozen in," and would be retained if the field were subsequently removed while the temperature was kept unchanged. By preparing different specimens in this way, it would be possible to produce a variety (in principle, an infinite variety) of different states, all existing under the same

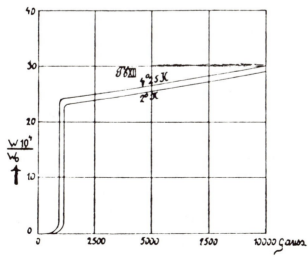

Fig. 4-6. Resistance of a lead coil as a function of applied magnetic field, for two values of temperature. [*Commun. Phys. Lab. Univ. Leiden* No. 139 (1913), p. 67, Fig. 1.]

external conditions, and presumably even in thermal contact with one another so that they should be in equilibrium. Until 1933, no experimental evidence conflicted with this idea, and a few experiments seemed to suggest it. There were even some theoretical treatments supporting it.

At this point, Meissner, studying the superconducting transition, was struck by the appearance of a sort of hysteresis effect: The return to the normal state for a single crystal of tin took place at a slightly higher temperature than the transition to the superconducting state. This effect was noticeable even when the individual points were measured with a current-reversal method designed to eliminate thermoelectric effects; it was heightened when the current was not reversed. This hysteresis suggested to him the likelihood of an alteration in the permeability of the specimen. As he put it, "If the distribution of the measuring current and its magnetic field were not altered, there was no basis for the appearance of hysteresis phenomena." He therefore undertook with his coworkers to determine the distribution of the field in the vicinity of the superconductor. He summarized the results this way: "If we have a cylindrical superconductor, as long as possible, which is represented in cross section in [Fig. 4-7], and a homogeneous magnetic field . . . directed perpendicular to the axis of the cylinder, then above the transition point the lines of force pass through the superconductor almost unhindered, corresponding to [part (a) of Fig. 4-7], since its permeability is approximately 1. . . . If we reduce the temperature

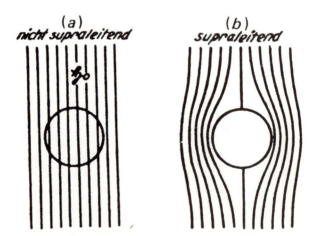

Fig. 4-7. Magnetic field distribution around a superconductor: (a) above the transition temperature, (b) below the transition temperature. [*Physik. Z.* **35** (1934), p. 933, Fig. 5.]

below the transition point, then the field distribution becomes approximately as is represented in [part (b) of Fig. 4-7].

"For these studies there was used a carefully made single crystal of tin of 10 mm diameter and 130 mm length. The magnetic field distribution was determined in the following way: Near the surface of the superconductor was a small coil of about 10 mm length and 1.5 mm breadth, the ends of which were connected to a ballistic galvanometer.[7] The coil could, firstly, be rotated around its long axis which was parallel to the axis of the superconductor; and, in addition, it could be carried around the axis of the superconductor on two different circles. One circle was chosen so that the coil lay close to the superconductor; the second, such that the coil was about 4.5 mm away from the superconductor. Now first for several points on both circles the direction of the magnetic field was determined by the method of turning the coil on its axis far enough that for a sudden rotation of the coil through 180° (again around its long axis which was parallel to the length of the superconductor), the ballistic galvanometer gave a maximum throw. Moreover, from this maximum deflection the magnitude of the magnetic field strength could be calculated. In this way the direction and magnitude of the magnetic field on the two circles was measured."

This simple description cloaks what must have been a laborious procedure. At each of (at least) sixteen points, eight on each circle, the coil was adjusted to a chosen angular position, then flipped through 180°, and the galvanometer throw was measured. Then the angular position of the coil was changed slightly, and the process repeated, to find the initial angle for which the flipping of the coil gave the maximum deflection of the galvanometer.

"In [Fig. 4-8] is shown, for various points on the two circles of 12 and 20 cm diameters, the direction of the plane of the coil for maximum ballistic throw and thus the direction perpendicular to the magnetic field; the points studied are designated by the values of the angle α shown in the figure." Figure 4-9 shows the actual field distribution implied by the directional distribution and the associated magnitudes measured as described above. "The directions $\alpha = 0$ and $\alpha = 180°$ correspond to the original direction of the homogeneous field H_0. The fact that the directions [in Fig. 4-8] for $\alpha = 0, 90, 180$, and 360° [*sic*; 270° is meant] are not parallel and perpendicular to H_0 arises partly from the fact that the circles of 12 and 20 mm diameters were not sufficiently concentric with the crystal axis. Partly, however, surely properties of the crystal, small crystal defects, or small deviations from complete homogeneity of the original field are to be held responsible for it."

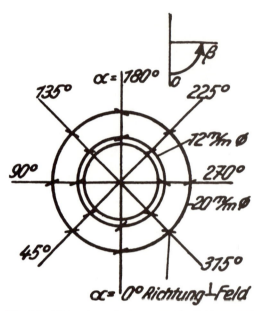

Fig. 4-8. "Direction of the magnetic field strength after passage below the transition point, according to Meissner and Heidenreich." The small marks show the plane of the flip coil for maximum galvanometer deflection and thus are perpendicular to the field directions. [*Physik. Z.* **35** (1934), p. 934, Fig. 6.]

These results alone could have been explained simply by the assumption that the permeability of a superconductor disappears below the transition point. This possibility was eliminated by the following modification.

". . . [T]he tin single crystal was provided with a hole of 6 mm diameter concentric with the axis of the cylinder. Then the magnetic field in the interior of this hole was determined above and below the transition point in a similar way to what was already described for the outside.

"The result is qualitatively shown first, in [Fig. 4-10]. The homogeneous field that was present in the interior above the transition point [part (a) of the figure] changed its direction a little (a few degrees) at the decrease below the transition point, but remained in actual existence. The rotation is shown greatly exaggerated in [part (b) of Fig. 4-10]. The field strength decreased by about 10% on the average. This can be seen more exactly in [Fig. 4-11]. . . .

"If the external field was now switched off, the magnetic field in the external vicinity of the tin tube vanished completely. In the interior, however, there remained the remanence field represented in

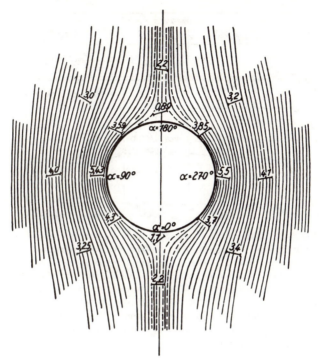

Fig. 4-9. Distribution of lines of force around a superconducting single crystal of tin with the directional indicators from Fig. 4-7 superimposed. [*Physik. Z.* **35** (1934), p. 934, Fig. 7.]

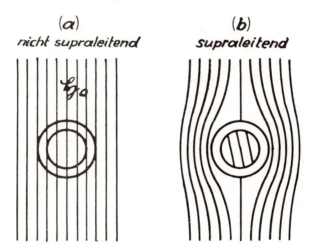

Fig. 4-10. Effect of transition on the distribution of magnetic field in the vicinity of a hollow cylindrical superconductor: (a) above the transition, (b) below the transition. [*Z. ges. Kälte-Ind.* **11** (1934), p. 127, Fig. 4.]

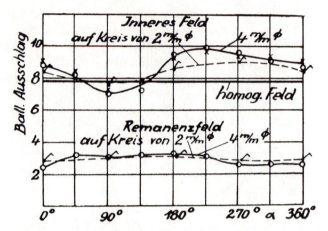

Fig. 4-11. Magnetic field strength inside the hole in the cylindrical supercon-ductor below the transition point. The angle α is measured from the original direction of the field. The upper curves are values determined before removal of the field: The heavy horizontal line is the strength of the original field; the solid curve with small circles and crosses, that on a circle of 4-mm diameter; and the dashed curve with flagged points, that on a circle of 2-mm diameter. The lower curves, indicated in the same manner as the upper curves, are the values obtained after removal of the external field. [*Z. ges. Kälte-Ind.* **11** (1934), p. 127, Fig. 5.]

the lower part of [Fig. 4-11]. Also this did not change in the course of time, but had the same magnitude after two hours. If the external field was now switched on again, there again resulted on the outside the field distribution [found earlier for the solid crystal]. In the interior, however, the state represented in the upper part of [Fig. 4-11] did not reappear, but rather the field strength increased only by 3 to 4% above the value of the remanence field given in [Fig. 4-11]. On switching off again, there appeared again the same remanence field. . . . After a hundred cycles of switching the external field on and off the remanence field was still completely unchanged."

". . . One sees at any rate from these studies on the distribution of a magnetic field in the interior of the hollow tin single crystal that the experimental results cannot be explained just by the assumption that the permeability drops to zero. For if this were generally the case, then also no field line could end on the inner surface of the hole in the superconductor, as is obviously the case according to the experimental findings."

It was still many years before a satisfactory theory on how superconductivity takes place was developed; in fact, the situation as of 1972 had not been entirely settled. But with Meissner's discovery, it became possible to give a satisfactory macroscopic, phenomenological treatment of the facts.

N

1. If   ed in the field and then made superconducting, the   ; if it is placed in the field after it becomes
supercondu

2. T   vapor rising from the liquid.

3. T   oule-Thomson or Joule-Kelvin effect, and is
utilized in   ling devices, such as household refrigerators.

4.   keep it from freezing.

5.   tial difference that Kamerlingh Onnes could
detect was 0.03 m

6. To a good approximation, the critical field at any temperature T below
the critical value T_c is given by $H_c(T) = H_c(0)(1 - T/T_c)^2$.

7. This is a galvanometer whose moving element has a large rotational
inertia, so that a short pulse of current can have completely passed through it
before it has moved appreciably. The effect is to produce a deflection essentially proportional to the charge that has flowed through the coil. The charge,
in turn, is proportional to the strength of the magnetic field at the test coil,
provided the latter is "flipped" quickly.

Bibliography

Kamerlingh Onnes's research was originally reported at meetings
of the Mathematics and Physics section of the Royal Academy of
Science in Amsterdam and published in Dutch in its *Proceedings.*
These publications were translated into English and issued as *Communications from the Physical Laboratory at the University of
Leiden.* The liquefaction of helium is reported in No. 108 (1908).
The work on superconductivity appears in Nos. 119, 120, 122, and
124 (1911); 133 and 139 (1913); and 140 and 141 (1914). These
last two cover studies of the persistence of the current in a closed
superconducting circuit, not treated in this book.

Meissner's work appears only in German: W. Meissner and
R. Ochsenfeld, *Naturwissenschaften* **21**, 787 (1933); W. Meissner,
R. Ochsenfeld, and F. Heidenreich, *Zeitschrift fur die gesamte Kälte-
Industrie* **11**, 125 (1934); W. Meissner, *Physikalische Zeitschrift* **35**,
931 (1934).

A review by Kamerlingh Onnes of his work up to the discovery
of the field effect, and the original letter of Meissner and Ochsenfeld,
are included among the works reprinted in *Superconductivity*:
Selected Reprints (American Institute of Physics, New York, 1964),
which also contains an annotated bibliography of further readings.
A second bibliography, covering more recent developments, is given
by D. M. Ginsberg, *American Journal of Physics* **38**, 949 (1970).

Chapter 5

The Strange Behavior of Liquid Helium

As soon as the liquefaction of helium was accomplished (see Chap. 4), the liquid itself became an object of study. There was a prompt indication that helium was an unusual sort of substance: Kamerlingh Onnes himself found in 1911 that as the temperature was lowered, the density went through a maximum. By the late 1920s, there were several pieces of evidence that at a temperature of about 2.2 K the properties of liquid helium changed noticeably and that the low-temperature form, at least, was anomalous. Then, in the period from 1935 to 1938, four completely unique forms of behavior were discovered that provided essential clues for the understanding of the nature of liquid helium. This chapter describes the various developments, with particular emphasis on the last four.

Kamerlingh Onnes's original results on the density comprised only a few rather widely spaced points on a curve of density as a function of temperature. The curve extended to a temperature that was low enough, though barely, to show the existence of a maximum but not to locate it accurately. The first precision measurements were made in 1924 by Kamerlingh Onnes and J. D. A. Boks and are shown in Fig. 5-1. The point to be noticed especially is not that there is a maximum—after all, ordinary water has that property—but that it appears as a cusp rather than as a smooth curve. However, Kamerlingh Onnes and Boks may not have been entirely convinced of the reality of that; they commented, "It will be well to determine more points in the neighbourhood of the maximum."

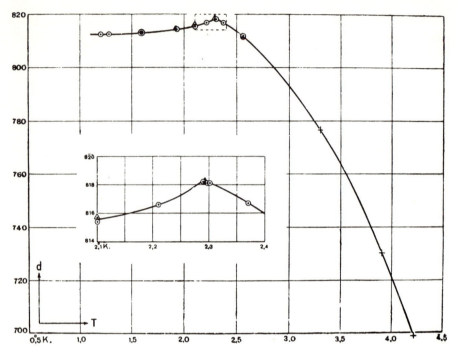

Fig. 5-1. The density of liquid helium as a function of temperature. The vertical axis is the ratio of the density of the liquid to that of (gaseous) helium at standard temperature and pressure (0°C and 760 Torr); the horizontal axis is the temperature in kelvins. [*Commun. Phys. Lab. Univ. Leiden* No. 170*b* (1924), p. 23, Fig. 5.]

The production of solid helium was finally achieved in 1924, by W. H. Keesom.[1] This does not mean that earlier attempts had not been made, however. Keesom says, "On the same day that Kamerlingh Onnes liquefied helium for the first time, he investigated whether it would become solid on further cooling by evaporation under reduced pressure. He was at that time able to obtain a vapour pressure of below 1 cm, probably 7 mm. The helium however remained liquid.

"This attempt to solidify helium by reducing the pressure under which it evaporates, was repeated on several occasions." In 1909, Kamerlingh Onnes was able to reach a temperature estimated at 1.4 K, and in 1910, 1.15 K. "But the helium still remained a thin liquid." Further attempts were "postponed in favour of more urgent questions," since "for the investigation of the solidification of helium new means were necessary."

"A new attempt was made in 1919, but with little improvement. Then in 1921, Kamerlingh Onnes by making use of a battery of condensation pumps, reached a remarkably lower pressure, . . . at which . . . the temperature should be $0°.82$ K. Meanwhile helium still remained liquid, so that Kamerlingh Onnes wondered whether helium would perhaps remain liquid even if it were cooled to the absolute zero. . . .

"Meanwhile . . . results . . . about the change of the melting point of hydrogen by pressure, had made me wonder, if it would not be possible at the temperatures already reached, to solidify helium by pressure." Keesom therefore constructed a device in which pressure was applied to both ends of a fine brass tube. The solidification of helium in the tube was shown by a difference in pressure between the two ends when one of them was opened, due to blockage by the solid. In this way, a rough curve was obtained. The measurements were then repeated by the same method with greater accuracy; and, it having been established that the pressure required was not extremely high (only of the order of several atmospheres, not several hundred), still again, this time using a glass vessel and registering solidification by observing visually the entrapment of a magnetically operated iron stirrer.

"In Fig. [5-]2 the melting curve is represented.

"The melting curve shows an important peculiarity: it bends at the lowest temperatures so as to become more and more parallel to the T-axis. . . . So unless the melting curve bends down again to the T-axis, . . . the surmise expressed by Kamerlingh Onnes . . . that helium (under its own saturation pressure) remains liquid down to the absolute zero . . . would be established.

"In the supposition mentioned, . . . coexistence between solid and gas is not possible. . . . This is yet another peculiarity, which distinguishes helium from all other substances."

Still another peculiarity is the apparent existence of a horizontal portion of the melting curve. By means of arguments from thermodynamics, it can be shown that this implies a vanishing latent heat of fusion, at least at absolute zero: The fusion is a purely mechanical and not a thermal process.

In 1927, Keesom and M. Wolfke, in measurements on the dielectric constant of liquid helium, "observed that at a temperature almost corresponding with that one at which Kamerlingh Onnes had found a maximum in the density curve, the dielectric constant shows a sudden jump." They noted that the density data "may be associated as well, if not better . . . with the admission of a jump . . . than with that of a smooth maximum"; and that at the same temperature peculiar variations occur in the specific heat, the heat of vaporization, and the surface tension. They therefore suggested "that at that temperature the liquid helium transforms itself into an other phase, liquid as well." They proposed the names "liquid helium I" for the

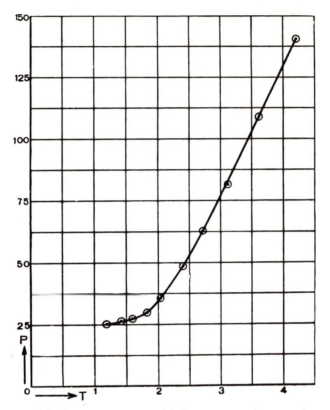

Fig. 5-2. The melting curve of helium: pressure in atmospheres versus temperature in kelvins. [*Commun. Phys. Lab. Univ. Leiden* No. 184*b* (1926), p. 17, Fig. 2.]

variety stable at the higher temperature, and "liquid helium II" for the variety stable at the lower temperature. They corroborated their suggestion by apparently observing the existence of a heat of transformation, evidenced by a kink in the curve of temperature versus time as the liquid helium warmed up.

This idea obtained still further, though qualified, confirmation in 1932, when Keesom and K. Clusius made more thorough measurements of the specific heat of liquid helium. The results they obtained, together with earlier measurements by L. I. Dana and Kamerlingh Onnes, are shown in Fig. 5-3. From the shape of the curve, somewhat resembling the Greek letter λ, the critical temperature has become known as the lambda point.[2] Particularly significant is the fact that while there is a discontinuity in the curve at 2.19 K, the curve does not become infinite, thus contradicting the idea of a heat of transformation. Keesom and Clusius checked this by supplying energy to the calorimeter in pulses and noting that in every instance the pulse

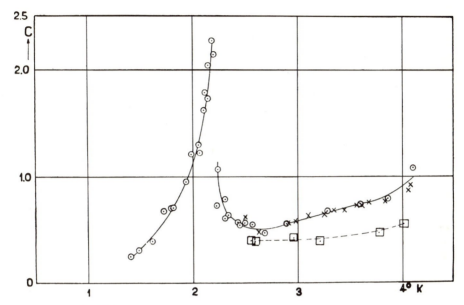

Fig. 5-3. The specific heat of liquid helium, in calories per gram per degree, as a function of temperature. The open points are the results obtained by Keesom and Clusius: squares for specific heat at constant volume, circles for that at saturated vapor pressure. The crosses represent earlier measurements of the specific heat at saturated vapor pressure by Dana and Kamerlingh Onnes. [*Commun. Phys. Lab. Univ. Leiden* No. 219e (1932), p. 51, Fig. 3.]

gave rise to a temperature increase—small, perhaps, especially in the neighborhood of 2.19 K, but always definitely nonzero. They say, however, ". . . [W]e might still call the observed effect a transformation from a state II into a state I."[3] They continue, "Although in the foregoing the nature of the transformation has been more nearly clarified, yet one must admit that its inner reason is still completely hidden from us." The best they could do was to associate it with similar transformations already known in certain solids, and in ferromagnetic substances at the Curie point.[4]

These various results were unexpected and admittedly perplexing; but, except for the absence of a solid-liquid-vapor triple point, none of them was really startling, and even that feature seems to have attracted no special attention.[5] The first truly exciting development was the discovery, in 1935, that liquid helium II appears to have an infinite thermal conductivity.

The discovery was made by W. H. Keesom and his daughter, Miss A. P. Keesom, in the course of further experiments on the specific heat. The method was to supply a "pulse" of heat to a calorimeter by means of an electric heater energized for a fixed time interval (usually 1 min) with a measured voltage and current, and to follow the temperature as a function of time. The temperature was measured by means of a resistance thermometer. The arrangement as used by Keesom and Clusius is shown in part (a) of Fig. 5-4, and its modified form as ultimately used by Keesom and Miss Keesom in part (b). Under ordinary circumstances, the thermometer would be "superheated" by conduction through the walls of the calorimeter and the copper cross and then settle to a new equilibrium value as the heat became distributed through the liquid helium.

Already in the earlier form of the apparatus, there were indica-

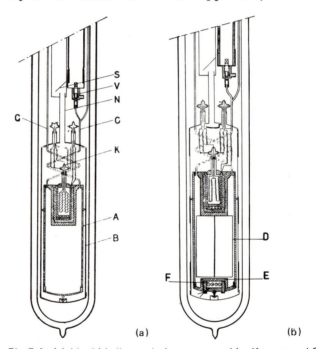

(a)   (b)

Fig. 5-4. (a) Liquid helium calorimeter as used by Keesom and Clusius and in the early work by the two Keesoms. *A* is the calorimeter cup, shielded from radiation by the vacuum-tight brass cylinder *B*; the space between can be evacuated or filled with gaseous helium for heat exchange, through the German silver tube *N*. The metal mirror *S* reduces the incidence of energy through *N* from above. *K* is a glass cross through which the thermometer leads are carried. *C, C* are capillaries; the one on the left connects through a valve to the high-pressure conduit of the helium compressor, while the one on the right splits into two branches, one leading to the manometer and the other to a shutoff valve *V* that forms part of the filling arrangement. The "core" carrying the heater and thermometer screws into the calorimeter vessel. The entire assembly is contained in a helium cryostat. [*Commun. Phys. Lab. Univ. Leiden* No. 219*e* (1932), p. 43, Fig. 1.] (b) The apparatus with the modifications effected by the Keesoms. *E* is the new core, carrying the heater *F; D* is a copper cross for improved heat transfer. All other parts are unchanged. [*Physica* **2** (1935), p. 559, Fig. 1.]

tions of a difference in the heating curves below and above the λ point. In the new version, ". . . it is clear that . . . with heating from below there is a marked difference between the after-periods below and above the lambda-point. Fig. [5-5]. . . shows this difference . . . markedly. It is very striking that the change in the shape of the after-period takes place at once in passing the lambda-point. We observe that for temperatures below the lambda-point the heat exchange in the calorimeter takes place again very rapidly, whereas above the lambda-point it takes some time."

They considered the possible effects of convection currents and were convinced that these could not account for the phenomenon. They concluded "that these experiments show that there is also a sudden change in the heat conductivity when passing from He II to He I," such that heat was conducted by He II far more effectively than it was through even the copper calorimeter vessel. They promptly set about investigating the change and reported the results in a brief paper in *Physica* in 1936.

"In our first experiments we examined the heat conduction in a layer of liquid helium which had the form of a circular cylinder, radius 18 mm, height 5 mm. The liquid helium layer was contained between two copper pieces each containing a heating unit and a phosphorbronze [resistance] thermometer. We . . . [obtained at

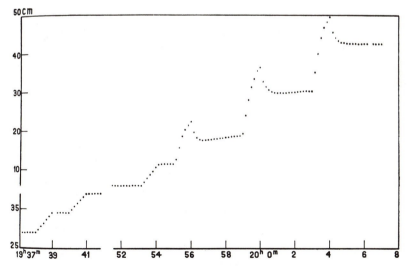

Fig. 5-5. The course of the temperature as a function of time for one of the series of measurements on the specific heat. (The vertical axis is graduated in centimeters of deflection of a galvanometer in the thermometer circuit.) For the three later (hence warmer) pulses, the overshoot of the thermometer is clearly evident, whereas it is undetectable in the earlier ones. [*Physica* **2** (1935), p. 569, Fig. 6.]

3.3 K a] value of the order of magnitude of the heat conductivity of gases at ordinary temperatures.

"At temperatures below the lambda-point the heat conductivity appeared to be too large to be measured in this way. When heating one copper piece the temperature of the other followed immediately without a measurable temperature difference. . . .

"For the next experiments we designed two copper pieces each containing a phosphorbronze thermometer and a liquid helium chamber of about 2 cm³. The two helium chambers were connected by a capillary of non-conducting material. One of the copper pieces was provided with a heating unit. From the helium chamber of the other a capillary started through which the whole could be filled with liquid helium. Of course the copper pieces and the connecting capillary were arranged in a space which could be evacuated. In these experiments it appeared that still the temperature of the two copper pieces followed each other too rapidly to make good measurements possible. . . .

"To eliminate the thermal resistance occurring at the limiting surface between copper and liquid helium II, and which appeared to be relatively very considerable, we made experiments with two different connecting capillaries, one long 22, the other long 94 cm. Internal diameter in each case 0.6 mm.

". . . We restrict ourselves in this preliminary report to the statement that we measured at 1.4 and 1.75°K values of the heat conductivity of liquid helium II of about 190 cal/degr. sec cm.

"We note that this heat conductivity is about 200 times that of copper at ordinary temperatures, or about 14 times that of very pure copper at liquid hydrogen temperatures. Hence liquid helium II is by far the best heat conducting substance we know.

"It appears further that the heat conductivity of liquid helium II at the temperatures mentioned is about 3×10^6 times that of liquid helium I.

"Connecting this with the abrupt change of the heat conductivity in passing the lambda-point we may perhaps be justified in calling liquid helium II *supra-heat-conducting.*"

The situation was not quite as simple as it seemed. Further studies on heat conduction by J. F. Allen, R. Peierls, and M. Zaki Uddin at Cambridge University in 1937 showed that the classical law of conduction does not hold, so that it is not clear how to define a coefficient of thermal conductivity. Their work did, however, confirm the Keesoms' conclusion that heat transfer in liquid helium II, whatever the mechanism, was far more effective than in any other known substance.

In the following year, the whole picture was to change dramatically. As Allen put it many years later, "The air was fairly humming then with new experiments and discoveries nearly every week for a period of two or three months." Allen himself had a role in two of the three principal discoveries.

The year opened with a Letter in *Nature* by Peter Kapitza, of the Institute for Physical Problems of the Academy of Sciences, Moscow, who had conceived of a possible explanation for the heat conductivity. An experiment to test the proposed explanation produced astonishing results.

"The abnormally high heat conductivity of helium II below the λ-point, as first observed by Keesom, suggested to me the possibility of an explanation in terms of convection currents. This explanation would require helium II to have an abnormally low viscosity; at present, the only viscosity measurements on liquid helium[6]... showed that there is a drop in viscosity below the λ-point by a factor of 3 compared with liquid helium at normal pressure, and by a factor of 8 compared with the value just above the λ-point."

In these experiments, however, no check was made to ensure that the motion was laminar and not turbulent.

"The important fact that liquid helium has a specific density ρ of about 0.15, not very different from that of an ordinary fluid, while its viscosity μ is very small comparable to that of a gas, makes its kinematic viscosity $\nu = \mu/\rho$ extraordinarily small. Consequently when the liquid is in motion in an ordinary viscosimeter, the Reynolds number[7] may become very high, while in order to keep the motion laminar, ... the Reynolds number must be kept very low. This requirement was not fulfilled in the ... experiments [referred to earlier], and the deduced value of viscosity thus ... may be higher by any amount than the real value.

"The very small kinematic viscosity of liquid helium II thus makes it difficult to measure the viscosity. In an attempt to get laminar motion the following method (shown diagrammatically in [Fig. 5-6]) was devised. The viscosity was measured by the pressure drop when the liquid flows through the gap between the disks 1 and 2; these disks were of glass and were optically flat, the gap between them being adjustable by mica distance pieces. The upper disk, 1, was 3 cm. in diameter with a central hole of 1.5 cm. diameter, over which a glass tube (3) was fixed. Lowering and raising this plunger in the liquid helium by means of the thread (4), the level of the liquid column in the tube 3 could be set above or below the level (5) of the liquid in the surrounding Dewar flask. The amount of flow and the pressure were deduced from the difference of the two levels, which was measured by cathetometer.

"The results of the measurements were rather striking. When there were no distance pieces between the disks, and the plates 1 and 2 were brought into contact (by observation of optical fringes, their

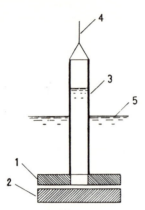

Fig. 5-6. Diagram of Kapitza's method of measuring the viscosity of liquid helium. [*Nature* **141** (1938), p. 74, unnumbered figure.]

separation was estimated to be about half a micron), the flow of liquid above the λ-point could be only just detected over several minutes, while below the λ-point the liquid helium flowed quite easily, and the level in the tube 3 settled down in a few seconds. From the measurements we can conclude that the viscosity of helium II is at least 1,500 times smaller than that of helium I at normal pressure."

Certain features of the flow suggested that it was actually turbulent, a suggestion borne out by the Reynolds number obtained from a value of viscosity calculated as if it were laminar. Here again, then, the viscosity is only an upper limit.

"... [B]ut the present upper limit (namely, 10^{-9} C.G.S.) is already very striking, since it is more than 10^4 times smaller than that of hydrogen gas (previously thought to be the fluid of least viscosity). The present limit is perhaps sufficient to suggest, by analogy with supraconductors, that helium below the λ-point enters a special state which might be called a 'superfluid'."

In the light of later results, to be described in the material that follows, Kapitza's results might have been called into question if they had been isolated. As it happened, however, similar results had been obtained at Cambridge University by Allen and A. D. Misener, using a quite different method. Their results were also reported in *Nature*, in a Letter immediately following that of Kapitza, and were described in detail, together with more extensive measurements, in the *Proceedings of the Royal Society* the following year.

"The apparatus and the experimental method employed were essentially the same for all of the flow measurements. A glass capillary was connected at one end to a cylindrical glass reservoir of such a diameter that the flow caused the liquid level in it to change at a rate convenient for measurement. The other end of the capillary was free and opened into the bath of liquid helium. The top of the reservoir was usually left open, but the opening was constricted to a diameter of from 0.5 to 1 mm., depending on the size of the reservoir. The purpose of the constriction was to reduce as far as possible any transfer of liquid to the reservoir by means of mobile surface films passing over the rim of the reservoir."

The last sentence refers to a discovery, made between the time of the initial note and that of the detailed report and to be discussed later, that a film of liquid helium II formed on the walls of a container and could transport the liquid over the top of the wall if the liquid levels on the two sides were unequal. It was this that might have qualified Kapitza's results.

"Most of the reservoirs were afterwards tested with the capillaries closed off, to measure the actual amount of liquid transferred by the films. The proportion thus transferred was generally less than 1%, although in the case of the finest glass capillaries it did amount to 7 or 8%.

"During the experiments the reservoir was always partially immersed in the bath. The flow was initiated by raising or lowering the reservoir and capillary, thereby separating the reservoir and bath levels. The rate of change of both reservoir and bath levels was then observed with a cathetometer. The latter was provided with a graduated eyepiece scale, and the times at which the levels passed successive divisions was recorded. . . .

"When long capillaries were used they were coiled horizontally to fit the flask containing the liquid helium bath. The diameter of the coils was usually 4 or 5 cm., i.e. approximately one hundred times the diameter of the capillary, which could, therefore, be considered as essentially straight. . . .

"Steady vapour pressures and hence constant temperatures were maintained by a manually adjusted needle value and a differential oil manometer in the helium pumping line. It was found to be essential to keep the temperature drift below 10^{-3} degree/min. during the measurements, particularly when they were carried out close to the λ-point.

"The maximum hydrostatic pressure used in the flow measurements was 15 mm. of liquid helium, and the cathetometer magnification was adjusted so that the 15 mm. of height occupied the whole field of view of the eyepiece.

". . . In all cases the change in reservoir level was plotted against the time. The slope of the graph at any point could then be transformed into the volume per second or the average velocity, v, of the liquid passing through the capillary. At the same point the difference between reservoir and bath levels gave the pressure head, p, corresponding to the velocity so determined. From each graph a velocity-pressure table was constructed which was plotted logarithmically. . . . The slope ($d \log v/d \log p$) of the logarithmic graph then represents the index of p in the velocity-pressure relation.[8] According to classical theory the slope or the index is unity for pure laminar flow, and is ½ for pure turbulent flow."

There were four parameters subject to variation—the length of the capillary, its radius, the pressure, and the temperature—and two terms in which results could be expressed, these being the velocity of flow and the slope of the logarithmic velocity-pressure graph. The extended article treats a variety of combinations of these. The crucial features, however, were already evident in the short note, which dealt with only two tubes, one with a circular cross section of radius 0.05 cm and length 130 cm, and the other with an elliptical cross section with semiaxes 0.001 and 0.002 cm and of length 93.5 cm.

". . . The data showing velocities of flow through the capillary and the corresponding pressure difference at the ends of the capillary are . . . plotted on a logarithmic scale in [Fig. 5-7].

"The following facts are evident:

"(a) The velocity of flow, q, changes only slightly for large changes in pressure head. For the smaller capillary, the relation is approximately $p \propto q^6$, but at the lowest velocities an even higher power seems indicated.

"(b) The velocity of flow, for given pressure head and temperature, changes only slightly with a change of cross-section area of the order of 10^3. . . .

"If, for the purpose of calculating a possible upper limit to the viscosity, we assume the formula for laminar flow, that is, $p \propto q$,[9] we obtain the value $\eta = 4 \times 10^{-9}$ C.G.S. units. . . .

"The observed type of flow, however, in which the velocity becomes almost independent of pressure, most certainly cannot be treated as laminar or even as ordinary turbulent flow. Consequently any known formula cannot, from our data, give a value of the "viscosity' which would have much meaning."

Meanwhile at Leiden, a research fellow from the United States, G. E. MacWood, and Keesom were measuring the viscosity of liquid helium by the classical method, that of observing the damping of rotational oscillations of a disk immersed in the fluid. The apparatus was so designed that either liquid or gaseous helium could be used, and the latter served as a calibration. Care was taken to see that the fluid motion was indeed laminar, as indicated by an accurately linear variation of the logarithm of the amplitude of oscillation with time. The results were not completely what might have been expected—for

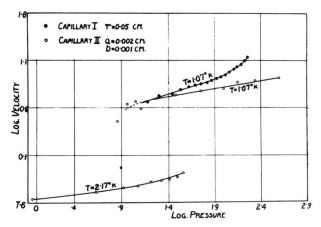

Fig. 5-7. Logarithmic plot of velocity as a function of pressure for flow of liquid helium II in capillaries. The velocities are measured in centimeters per second, the pressures in dynes per square centimeter. [*Nature* **141** (1938), p. 75, unnumbered figure.]

example, the viscosity was found to increase with increasing temperature, in contrast to the behavior for normal liquids—but they were certainly far from the values implied by the work of Kapitza and that of Allen and Misener. Allen visited Leiden, and recalls how he and MacWood "argued over the desk top for quite some time during two or three days . . . as to how one could reconcile the two totally conflicting results on classical decay of amplitude of an oscillating disc and the Poiseuille flow through a long capillary. In the end of course we were both right, but at the time the whole thing was very mysterious since both of us believed implicitly in our own results."

The next discovery, which involved the most spectacular phenomenon, was reported by Allen and H. Jones in a Letter in *Nature* just four weeks later (and treated in detail in an Article by Allen and J. Reekie in the *Proceedings of the Cambridge Philosophical Society*). The discovery was made in the course of an extension of the work of Allen, Peierls, and Zaki Uddin on heat conductivity (see p. 64).

"It was noted that the conductivity appeared to increase as the bore of the conduction capillary decreased. It was then found that at very low temperatures (1.08°K.) small heat flows produced a rise in the level of the liquid in the closed bulb at the heated end of the capillary, instead of the fall caused by increased vapour pressure which was observed for higher heat flows. The apparatus was modified (Fig. [5-8]) so that the top of the bulb at the hot end of the capillary was open to the vapour above the surface of the liquid helium bath.

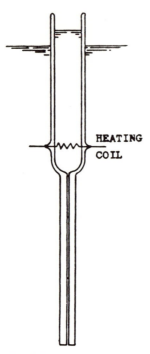

HEATING COIL

Fig. 5-8. Diagram of the apparatus with which Allen and Jones discovered the pressure effects associated with heat conduction in liquid helium [*Nature* **141** (1938), p. 243, Fig. 1.]

There was thus no possibility of a difference in vapour pressure between the liquid at either end of the capillary. As in the previous cases, heat was introduced electrically by means of a coil inside the bulb. On heating, the level of the liquid again rose in the bulb. The rise in level increased with increasing heat flow, and for constant heat flow increased with decreasing temperature. . . .

"A more striking manifestation of the above effect was observed by one of us (J. F. A.) in collaboration with A. D. Misener. Observations were being made on the flow of liquid helium II through a tube packed with fine emery powder (Fig. [5-9]). The top of the tube was allowed to project several centimetres above the level of the helium bath, and an electric pocket torch was flashed on the lower part of the tube containing the powder. A steady stream of liquid helium was observed to flow out of the top of the tube as long as the powder was irradiated.

". . . The apparatus forms, in effect, a very simple and efficient helium pump. The height of the jet becomes greater for lower bath temperatures and for powder particles of smaller size. A photograph

Fig. 5-9. Diagram of the apparatus with which the fountain effect is displayed. [*Nature* **141** (1938), p. 243, Fig. 2.]

of the liquid helium pump is shown in Fig. [5-10]. The stream rises approximately 4 cm. above the jet, the top of the stream striking the side wall of the flask. . . . The maximum height of jet produced was 16 cm."

Evidently there was an intimate relationship between heat flow and mass flow in helium II. The nature of this relationship was to be made more explicit as a result of studies on the fourth (and last) of the "super" properties, that of film flow.

In contrast to the other three aspects of superfluid behavior, film flow had been hinted at much earlier.[10] Kamerlingh Onnes had noted as early as 1922 that the levels of liquid helium in two concentric containers tended to become equal; he attributed the behavior to some sort of "distillation" phenomenon. In 1936, Nicholas Kürti, B. V. Rollin, and F. Simon reported in an Article in *Physica* dealing with experiments on cooling by adiabatic demagnetization

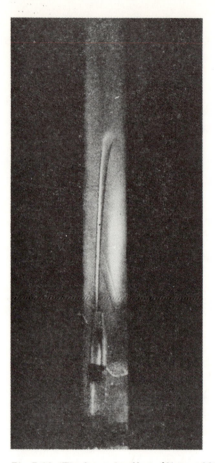

Fig. 5-10. The fountain effect. [*Nature* **141** (1938), p. 243, Fig. 3.]

that "... vessels containing liquid helium cooled below the λ-point and connected by a tube to the warmer parts of the apparatus show an abnormally high heat in-flow. This behaviour can probably be explained by the existence of a liquid layer along the walls of the tube." It was Rollin, reporting similar effects in more detail at the Seventh International Congress of Refrigeration, who suggested the relation to Kamerlingh Onnes's observation. Accordingly, at Oxford John G. Daunt and Kurt Mendelssohn set about investigating the phenomenon more carefully. They reported their results in a Letter in *Nature* in 1938 and described the work in detail in two successive Articles in the *Proceedings of the Royal Society* in 1939.

"In one very convenient apparatus [see Fig. 5-11] we utilized a small glass beaker suspended on a thin glass fiber which could be

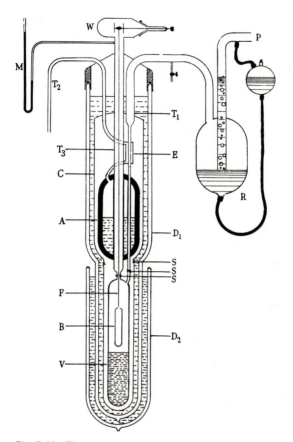

Fig. 5-11. The cryostat used by Daunt and Mendelssohn, as arranged for raising and lowering a beaker. The description is given in the first of the two Articles: "The vessel *V* in which the experiments were carried out is attached to a small helium liquefier.... In those experiments where a beaker *B* was employed this could be lifted and lowered by a thread or wire *F* which was carried by the tube T_3 out of the apparatus. The upper end of *F* was connected to a small vacuum tight winch *W* or a magnetic device.... The whole arrangement is included in a vacuum case *C* which is surrounded by liquid hydrogen in the Dewar vessel D_1. The lower end of the apparatus is made out of glass and connected to the main part by the copper-glass seals *S*. The tail of D_1 is left unsilvered and dips into a bath of liquid nitrogen in the Dewar vessel D_2." [*Proc. Royal Soc., Ser. A* **170** (1939), p. 425, Fig. 1.]

lowered into a bath of liquid helium II or raised out of it. . . . The following phenomena were observed:

"(a) When the empty beaker was lowered into the liquid, it filled up to the level of the bath although the rim of the beaker was everywhere above the level of the liquid.

"(b) When the beaker was partly lifted out of the bath, the level of the liquid in the beaker fell at the same rate as it rose in (a), and the level of the bath rose until both had reached the same height.

"(c) In order to establish whether the transfer took place by distillation through the gas phase or by transfer over the surface, the effect was examined of introducing 'wicks' of twisted copper wire which increased the surface leading from one level to the other. It was found that this increased the rate of transfer to several times the previous rate so long as, but only so long as, the wick reached into the liquid at the higher level [see Fig. 5-12].[11]

"(d) When the beaker (without a wick) was lifted completely out of the bath, it was found that the liquid vanished at the same rate as when it was still partly dipping into it. This was accounted for when it was observed that the liquid collected in drops at the bottom of the beaker and dripped into the bath.

"(e) The rate of transfer did not appear to differ very greatly whether the beaker was almost full or nearly empty, the rate decreasing only by 20 percent when the level had dropped from within 0.5 mm. of the rim to some 20 mm. from the rim."

Indeed, as shown in the first of the two Articles, the flow rate seemed to be almost completely independent of the relative levels.

"In order to investigate this question more in detail the transfer from a long beaker was measured as a function of the time. The beaker was filled by dipping it entirely into the bath and lifting it out again. Readings of the height of the level in the beaker were taken every 2 min. The level of the bath was kept at the same height in respect to the beaker during the experiments by adjusting the height of the beaker. The height of both levels is plotted against time in Fig. [5-13], showing the results of a typical experiment.

"The level inside the beaker drops quickly during the first 6 min. until it reaches a height of 1 cm. below the rim. From minute 6 onwards the height of the inner level is a linear function of time and height; this means that the velocity of transfer is constant and does not depend on the difference in height between the inside and outside levels. In order to test this point at minute 33 the difference between the two levels was suddenly decreased by 65%. No change however was effected by this means in the velocity of transfer, as can be seen clearly from fig. [5-13]."

(Actually, more precise measurements did indicate an effect, although it was very small, barely outside the accuracy of the measurements, and far from what would be found for a siphon effect.)

"As the difference in height between the two levels does not influence the flow appreciably it was thought that perhaps the dis-

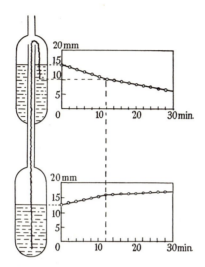

Fig. 5-12. Method of showing the effect of a copper "wick" on the film flow—left, a diagram of the apparatus; right, graphs of the time variation of the liquid levels in the two vessels. Note the change of slope in both curves as the level in the upper vessel drops below the end of the wick. [*Proc. Royal Soc., Ser. A* **170** (1939), p. 428, Fig. 3.]

tance the atoms in the film had to travel might be of importance. . . . In consequence we made an experiment with a beaker in which the path along one wall was very much longer than on the other (fig. [5-14]). The transfer is again recorded as difference in height of level in the beaker and in the bath plotted against time. For the first 30 min. the level of the bath was kept constant. . . . After 30 min. the level of the bath was raised so that it stood above the lower end of the inner tubes. The transfer velocity did not change but it was observed that now drops of liquid helium fell off the lower end of the innermost tube. At minute 45 the level of the bath was lowered slightly below the level in the beaker and it can be seen that, although the path on the inner wall is much longer, the flow is in the reverse direction and we observe exactly the same velocity of transfer as when helium flowed into the beaker.

"An interesting feature in the last experiment is the formation of drops observed when the lower end of the tubes was below the bath level. These drops seemed to indicate that the free liquid can be formed out of the film at such places which are below the upper level.[12] In order to investigate this further a beaker with a constriction in the inner wall was employed (fig. [5-15]). The empty beaker was dipped into the bath so that the level of the latter stood well above

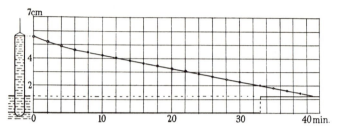

Fig. 5-13. The rate of change of liquid level in a beaker of liquid helium II. [*Proc. Royal Soc., Ser. A* **170** (1939), p. 430, Fig. 5.]

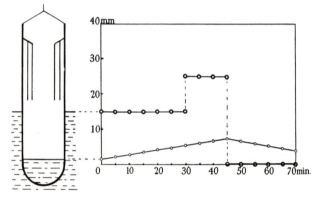

Fig. 5-14. Experiment to examine effect of path length on liquid transfer by the film—single circles, inner level; double circles, outer level. [*Proc. Royal Soc., Ser. A* **170** (1939), p. 436, Fig. 9.]

the height of the constriction. Then the beaker filled up and the level of the bath fell accordingly, the velocity of transfer being constant until, after 15 min., the bath level had dropped to the height of the constriction. At this time a sharp change in the transfer velocity is observed. From now on the decrease of the bath level and the corresponding increase in the level of the beaker is much smaller.

"... This means: the transfer between levels of liquid helium is limited by the narrowest place in the connecting surface *above* the heigher level."

The Letter in *Nature* which followed that of Daunt and Mendelssohn was by A. K. Kikoin and B. G. Lasarew of the Ukrainian Physicotechnical Institute (Kharkov), who had also investigated the film but with a quite different technique.

"A glass tube 4 mm. in diameter, insulated by a vacuum vessel, is immersed in liquid helium with only its lower part surrounded by the liquid; and the remainder is allowed to project above the level of the liquid helium bath. On the upper parts of the glass tube a copper

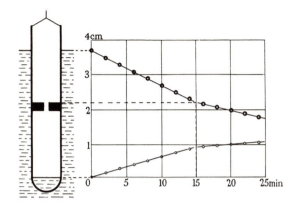

Fig. 5-15. Experiment showing the effect of a constriction in the beaker wall—single circles, inner level; double circles, outer level. [*Proc. Royal Soc., Ser. A* **170** (1939), p. 437, Fig. 10.]

tube is soldered, on to which a heating coil of constantin wire (0.1 mm.) and a resistance thermometer of phosphorbronze (0.1 mm.) are wound.

"When the lower part of the tube is surrounded by liquid helium II, its upper part has the same temperature as the lower part, that is, the temperature of the liquid helium in the Dewar vessel. When heat was introduced electrically by means of the heating coil, almost no heating is observed. . . . At a fixed value of the current, intensive heating begins. . . . When the current in the heating coil is switched off, the upper part of the tube very rapidly (in a few seconds) acquires the same temperature as that of the lower part. . . .

"It appears that over the surface of the tube a fine film of liquid helium II is formed, which, owing to its abnormally high heat conductivity, equalizes the temperature of the upper and lower parts of the tube. . . . This film at the limiting value of the current evaporates, and only then does the heating begin."

Kikoin and Lasarew estimated the thickness of the film as being on the order of 10^{-5} cm [in a Letter later in the year, they reported actual measurements of the thickness, giving $(2-3) \times 10^{-6}$ cm], and pointed out that convection currents could scarcely be set up in such thin films. Consequently, the heat transfer must have been due, they concluded, to the thermal superconductivity of helium II. Daunt and Mendelssohn, however, having discovered the actual mass transfer by the film, examined the question more closely, as reported in the second of their Articles.

"We worked with a beaker arrangement similar to that used in the previous experiments, employing a small Dewar vessel as beaker

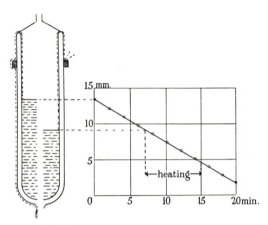

Fig. 5-16. Daunt and Mendelssohn's arrangement for determining the effect of thermal conduction through the surface film. [*Proc. Royal Soc., Ser. A* **170** (1939), p. 441, Fig. 1.]

(fig. [5-16]). This vessel bore on its outside a copper ring in which heat could be generated by an induced current. The beaker was first partly filled with liquid helium and then entirely lifted out of the bath. Owing to the transfer, the level in the beaker fell steadily and liquid collecting at the outside bottom surface of the Dewar vessel fell in drops into the bath. Now heat was supplied to the film at the outside of the beaker by inducing a current in the ring. It is evident that if the film 'conducts' heat, this heat will be carried back into the beaker and result in an additional evaporation of helium there. This means that during the heating period the liquid level in the beaker should fall more rapidly than before. Fig. [5-15] shows clearly that no change in the rate of flow took place when the ring was heated (minute 7-15). In the experiment represented in this figure so much heat was supplied that the film was entirely evaporated, which was also clearly shown by the disappearance of the drops. In other experiments the film was only partly evaporated; the amount of heat supplied was checked by the change in the number of drops falling off the beaker per min. These experiments were carried out at various temperatures, always yielding the same result. Further, it made no difference whether the level of the liquid in the beaker was above or below the level of the ring.

"These experiments show that no appreciable amount of heat has been conducted through the film from the heater into the beaker. . . . The heat transport in a tube containing liquid He II is thus a dynamic process. Contrary to the assumptions of Kikoin and Lasarew a convection of helium is actually responsible for the heat transport but this convection takes place in two phases. The transfer along the film carries the helium to the place of higher temperature from which it flows back as vapour, in a cycle somewhat similar to that described by water in the ocean currents and the atmosphere."

The phenomenon was summarized as follows:

"A container C (fig. [5-17]) is filled with liquid He II. The adjoining solid surface S is then covered with a helium film of about 3.5×10^{-6} cm. thickness. This film spreads over the whole surface until it either drops off as free liquid at a place below the level in C, or is evaporated. The rate at which the liquid disappears from C is the same whether the film drops off at A or is evaporated at B. It seems therefore that as soon as the film is removed from S by whatever means, the bared surface is readily covered again with helium from C. The 'rate of transfer' at which this regeneration of the film is carried out depends to a first approximation only on the temperature. This means that helium will always be transferred to any place of the surface from which the film has been removed."

By this time there were already some theoretical considerations in circulation. Fritz London had suggested that below the λ point, liquid helium consisted of two components: a normal liquid, and a set of atoms condensed into a single energy state, as is possible for particles such as helium atoms that obey Bose-Einstein statistics. In a Letter in *Nature* immediately following those of Daunt and Mendelssohn and of Kikoin and Lasarew, Laszlo Tisza examined some of the consequences of this scheme. He noted that the damping of an oscillating cylinder or disk would be effected only by the atoms of the normal liquid, but that the Bose-Einstein condensate would form a "superfluid" of essentially zero viscosity. This resolved, at least qualitatively, the discrepancy between the results of Keesom and MacWood (which had not been published at the time; in fact the published version referred to Tisza's note) and those of Kapitza and of Allen

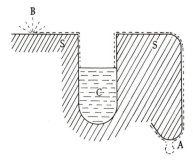

Fig. 5-17. Sketch to illustrate the summary discussion of film flow. [*Proc. Royal Soc., Ser. A* **170** (1939), p. 446, Fig. 5.]

and Misener. Tisza also showed that the two-fluid hypothesis would lead, first, to the production of a temperature gradient in a capillary as a result of flow of liquid helium II through it, and second, as an inverse process, to the fountain effect. In another Letter in *Nature* in 1938, Daunt and Mendelssohn reported an extension of their work on film transport that partly corroborated these ideas.

"(a) A small Dewar vessel (see Fig. [5-18]) containing a heating coil was suspended by a thread in a bath of liquid helium II. When no heat was supplied, the levels of the liquid both inside and outside the vessel adjusted themselves to the same height L_1, owing to the 'transfer' through the film on the interconnecting glass surface. When, however, a current was passed through the heating coil, the level of the liquid inside the vessel *rose* above the outside level and took up an equilibrium position L_2. . . . [D]ifferences between inside and outside levels of up to 5 mm. could be obtained. This clearly shows that there exists a 'transfer' of helium from a colder to a hotter place when a temperature gradient is imposed. . . .

"This effect is quite analogous to the 'fountain phenomenon' in the bulk liquid, discovered by Allen and Jones. . . .

"(b) A Dewar vessel (see Fig. [5-19]) was closed at the top and had a hole at the lower end which was constricted by a plug, P, of fine emery powder. It contained a phosphorbronze thermometer, T, and was suspended in a bath of liquid helium II. When the Dewar vessel was lifted out of the bath, the liquid ran rapidly out of the

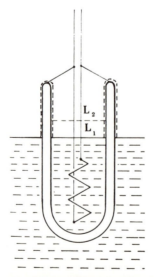

Fig. 5-18. Apparatus used to demonstrate flow of liquid helium II against a temperature gradient. [*Nature* **143** (1939), p. 719, Fig. 1.]

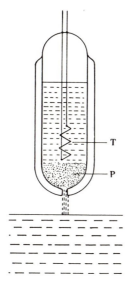

Fig. 5-19. Apparatus for demonstrating the "mechano-caloric" or inverse fountain effect. [*Nature* **143** (1939), p. 719, Fig. 2.]

vessel through P and fell into the bath, and at the same time the temperature of the inside liquid was noticed to rise by about 0.01°. On lowering the vessel so that now liquid ran from the bath into the vessel, the liquid inside was cooled by a similar amount.

"This mechano-caloric effect is evidently the reverse of the 'fountain phenomenon'. . . . Such a caloric effect has actually been postulated by Tisza for a flow of liquid helium II through capillaries. It seems to us, however, that the anomalous phenomena of liquid helium II are not so much caused by capillary flow as by a transport along solid surfaces; and these results seem to indicate that the heat content of these atoms transported by surface flow must be lower than average."[13]

There were still many phenomena to be discovered and many theoretical developments to be worked out. For example, Kürti and Simon reported in 1938 that the thermal superconductivity seemed to disappear below about 0.5 K. The work was interrupted, and resumed after World War II by H. A. Fairbank and John Wilks, by which time such behavior had been foreseen as a consequence of the kinds of excitations possible in a superfluid. As another example, experiments in 1973 by D. M. Lee and coworkers at Cornell University suggested that ^{3}He may also show superfluid properties, even though it is subject to Fermi-Dirac rather than Bose-Einstein statistics.[14] But the basic ideas had been established in the span of four years recorded in this chapter.

Notes

1. Keesom became codirector of the Physical Laboratory of the University of Leiden after Kamerlingh Onnes's retirement in 1923.

2. This term was introduced by W. H. Keesom and Miss A. P. Keesom at the suggestion of P. Ehrenfest.

3. Such transformations were not at that time regarded as phase transitions. They have since been given the name "phase transitions of the second kind," while those that involve a heat of transformation are called "of the first kind."

4. Ferromagnetism in any substance disappears above a temperature characteristic of the substance, called the "Curie temperature."

5. There was some criticism of the suggested course of the melting curve below the lowest temperature achieved, 1 K; but the alternative proposal was that it would turn upward, not downward. The implication that the fusion process is of a purely mechanical nature seems not to have been noted either by Keesom or by his critics, probably because the data did not extend to temperatures low enough to make the point inescapable. In fact, no measurements below 1 K were made until 1963, and measurements of sufficient precision to trace a melting curve were first made in 1966. It is now known that there is a minimum, at 0.775 K, but it is extremely shallow; at 0.3 K, the curve appears to be leveling off to a value only 8×10^{-3} atm above the minimum.

6. These measurements were made by a standard method, that of observing the damping of rotational oscillations of a cylinder.

7. The *Reynolds number* is defined as vl/ν, where v is the speed of the fluid motion, l is a characteristic dimension of the channel in which it is confined, and ν is the kinematic viscosity as given in the text.

8. This means the value of n if the pressure and velocity are related as $v \propto p^n$.

9. The classical law for laminar flow would also give a velocity proportional to the cross-sectional area of the tube.

10. Actually, there had been at least one early indication of abnormally low viscosity: In 1930 Keesom and J. N. van den Ende found that a vessel which showed essentially no leakage above the λ-point temperature was quite leaky below it. There seems to have been no follow-up on this point, however.

11. The possibility of capillary action between the wires of the wick was ruled out by replacing it with several single wires.

12. Compare the conditions under which drops were formed in these two cases (p. 74).

13. Tisza's ideas had already been used by Daunt and Mendelssohn to explain the initial behavior of flow curves such as those shown in Fig. 5-13.

14. In Fermi-Dirac statistics, a given quantum state can be occupied by no more than one particle of a given kind.

Bibliography

The background up through 1932 is summarized in F. London, *Superfluids*, vol. 2 (Wiley, New York, 1954), pp. 1-17.

Thermal superconductivity: W. H. Keesom and A. P. Keesom, *Physica* **2**, 557 (1935), and **3**, 359 (1936). See also J. F. Allen, R. Peierls, and M. Zaki Uddin, *Nature* **140**, 62 (1937).

Superflow: P. Kapitza, *Nature* **141**, 74 (1938); J. F. Allen and A. D. Misener, *Nature* **141**, 75 (1938), and *Proceedings of the Royal Society, Ser. A* **172**, 467 (1939).

Fountain effect: J. F. Allen and H. Jones, *Nature* **141**, 243 (1938); J. F. Allen and J. Reekie, *Proceedings of the Cambridge Philosophical Society* **35**, 114 (1939).

Film transport: J. G. Daunt and K. Mendelssohn, *Nature* **141**, 911 (1938), **142**, 475 (1938), and **143**, 719 (1939), and *Proceedings of the Royal Society, Ser. A* **170**, 423, 439 (1939); A. K. Kikoin and B. G. Lasarew, *Nature* **141**, 912 (1938), and **142**, 289 (1938). For anticipatory work in this direction, see N. Kürti, B. V. Rollin, and F. Simon, *Physica* **3**, 266 (1936), especially footnote ** on page 269; B. V. Rollin, in *Actes du Septième Congrès International du Froid*, vol. I (J. van Boekhoven, Utrecht, the Netherlands, 1937) p. 187. See also B. V. Rollin and F. Simon, *Physica* **6**, 219 (1939).

Chapter 6

Precision Values for Nuclear Magnetic Moments

Once the validity of Rutherford's nuclear model of the atom was confirmed by Geiger and Marsden's experiments[1] and supported by Bohr's theory of atomic structure, it was natural for the nucleus itself to be an object of attention. Of course, little was known of its composition, and even less about the forces that might bind the constituents. But an even more significant hindrance was that for a long time there were very few data that would reflect the detailed structure and behavior of the nucleus.

Some progress was made during the 1920s. The development of the "new" quantum theory provided at least a conceptual framework in which to operate. Then in the 1930s, the field of nuclear physics burst open, with the development of particle accelerators and the discovery of the neutron providing far more effective probes than had previously been available. Models of the nucleus began to be constructed in the hope of unifying the masses of reaction data that began to accumulate.

Meanwhile, there were other data that these models had to accommodate. The nucleus is an electromechanical system and therefore should have electromagnetic moments. In fact, some values of magnetic moments had been deduced as early as 1932. These values were comparatively crude, however, and more precise values were needed. I. I. Rabi and coworkers, in introducing a high-precision method of measuring magnetic moments, put it this way: "The magnetic moment of the atomic nucleus is one of the few of its important properties which concern both phases of the nuclear problem, the nature of the nuclear forces and the appropriate nuclear model. According to current theories the anomalous moment of the proton is directly connected with the processes from which nuclear forces arise. The question whether the intrinsic moments of the proton and neutron are maintained within the nucleus is part of the problem of two and multiparticle forces between nuclear constituents. With regard to the atomic model it is clear that the nuclear angular momentum does not alone suffice to fix the nature of the wave functions which specify the state of the nucleus. The magnetic moment, on the other hand, is sensitive to the relative contributions of spin and orbital moment and, with the advance of mathematical technique, suffices to decide between the different proposed configurations.

"In the light of these considerations it is particularly desirable that nuclear moments be known to high precision because small effects may be of great importance." These points were cogent enough that the new method and its applications won Rabi the Nobel prize for physics in 1944. This chapter describes how the method was developed and how it works.

The concept of a nuclear magnetic moment dates back at least to 1915, when H. S. Allen suggested that the components of the nucleus—then thought to be protons and electrons and perhaps α particles—would presumably be in some sort of rotational motion and, being charged, would therefore give rise to magnetic moments. The same sort of suggestion was made in 1921 by A. E. Oxley. Wolfgang Pauli, in 1924, suggested that the hyperfine structure[2] of atomic optical spectra resulted from an interaction between the orbital motion of the electrons and a net angular momentum of the nucleus; he did not feel it necessary to specify the mechanism of the interaction, apparently regarding the arguments of Allen and Oxley as self-evident. Pauli's ideas, however, were considered and explictly rejected by Georg Joos and by Arthur E. Ruark and R. L. Chenault in analyses of hyperfine structures (but Ruark had used an inappropriate value for the magnitude of nuclear moments in making his estimates). Similarly, H. Nagaoka, Y. Sugiura, and T. Mishima had suggested a sort of nuclear dynamics effect to explain the results of their measurements of hyperfine structure in mercury and bismuth; but their ideas were incompletely thought out and their data did not really justify their speculations, and they were ridiculed by Carl Runge, who claimed that the letters of Nagaoka's name could be related as well to his wavelength data as could any property of the nucleus.

In spite of all this, W. Wessel, in 1926, wrote, "The question of a possible magnetism of atomic nuclei has been raised several times, but as far as we know, has never been discussed in detail." Wessel's concern was the extent to which such magnetism might be expected to affect the scattering of α particles by nuclei; the answer was, not enough to be detectable.[3]   But the idea had taken deeper hold than Wessel recognized.   In 1927 and 1928, Ernst Back and Samuel Goudsmit, unaware of Pauli's suggestion but impressed by the finding of Joos and of Ruark and Chenault that some hyperfine structure patterns were miniature multiplets, made the first connection between hyperfine-structure measurements and nuclear moments and deduced the "spin" of the bismuth nucleus.[4]   Still later, in 1932, Gregory Breit and F. W. Doerman deduced from spectral measurements what appears to have been the first actual value for a nuclear magnetic moment, that of ^{7}Li.

The optical method has been extremely useful.   It is still the only source of data for many nuclear moments.   However, it is very demanding experimentally; and it requires a computation of the value of the electron wave function at the nucleus, which introduces considerable uncertainty into the result.   A more direct approach might thus enjoy significant advantages.   Indeed, Otto Stern, in the article in 1926 that initiated the remarkable series of "Studies with the Method of Molecular Rays from the Institute for Physical Chemistry of Hamburg University," pointed out that the method would be useful for measuring nuclear magnetic moments and even estimated the order of magnitude that would be expected for such moments: the "nuclear magneton," $e\hbar/m_N c$.

The molecular-beam method was not without its own difficulties.   In the third paper of the series, published in the same journal issue as the first, Friedrich Knauer and Otto Stern reported the successful detection and crude measurement of a magnetic moment of the appropriate magnitude, arising from the rotational motion of the hydrogen atoms in the water molecule. Their attempts to detect and measure a nuclear magnetic moment in mercury, on the other hand, were inconclusive.[5]   It was not until 1933 that Immanuel Estermann, Robert Frisch, and Stern succeeded in measuring a nuclear magnetic moment, that of hydrogen.   Their apparatus and method were basically the same as those used in the Stern-Gerlach experiment.[6]   To eliminate the effects of the atomic electrons, molecular hydrogen was used; this still left an effect due to the rotation of the molecule, which was measured empirically by studies on parahydrogen (the form of the molecule in which the nuclear spins are in opposite directions so that the molecular effect is the only one present) and used as a correction in the work with ordinary hydrogen.

Meanwhile Rabi, who had been a member of Stern's group in

1927 during a fellowship in Europe, had returned to Columbia University in 1929 and had begun a program of molecular-beam research there. Under his guidance, a series of modifications were made that culminated in what is known as the *molecular-beam resonance method*.[7] The first of these made use of the fact that in an atom with both an electronic and a nuclear magnetic moment, the effective total moment (defined as the derivative of the orientational energy with respect to the externally applied magnetic field H) depends on the magnitude of H, and also on the magnetic substate, that is, the component of the total angular momentum in the direction of H. The analysis will not be given here; the result is that the nature of the deflection pattern at a given value of H reveals the value of the nuclear magnetic moment μ_I.

Both this method and the original Stern-Gerlach deflection method have the drawback that the deflection depends on the speed of the atom or molecule. With a thermal beam, there is a spread of velocities and, consequently, a spread in deflections—a smearing of the pattern. If the nuclear angular momentum[8] is greater than $\frac{1}{2}$, however, there are certain values of H for which the effective moment vanishes and thus there is no deflection, and these values are independent on the atomic velocity. A measurement of the value(s) of H for which the intensity at the undeflected position is a maximum leads, therefore, to determination of both the nuclear angular momentum I and μ_I. This method has the further advantage that only H, and not its gradient, needs to be measured. An additional modification, which will not be further detailed, made it possible to work with nuclei for which $I = \frac{1}{2}$, still without the problem of velocity spread, but at the expense of again requiring knowledge of the field gradient.

The final refinement, which will be discussed in detail, was reported by Rabi, J. R. Zacharias, S. Millman, and P. Kusch, in two "Letters to the Editor" (1938) and an Article (1939), in *The Physical Review*. It was based on a concept suggested two years earlier by C. J. Gorter, which, roughly speaking, holds that a precessing gyroscope will absorb energy from a periodic perturbation only if the frequency of the perturbation is equal, or nearly so, to the frequency of precession. It "applies not only to nuclear magnetic moments but rather to any system which possesses angular momentum and a magnetic moment. We consider a system with angular momentum, J, in units of $h/2\pi$, and magnetic moment μ. In an external magnetic field H_0 the angular momentum will precess[9] with the Larmor frequency, ν, (in radians per sec.) given by,

$$\nu = \mu H_0/Jh. \qquad\qquad ([6\text{-}]1)$$

Our method consists in the measurement of ν in a known field H_0. The measurement of ν is the essential step in this method, since H_0 may be measured by conventional procedures. Using Eq. ([6-]1) we obtain the gyromagnetic ratio $[\mu/J = g]$. If, in addition, the angular momentum, J, of the system is known, we can evaluate the magnetic moment μ. In its present state of development our method is not suitable for the measurement of J.

"The process by which the precession frequency ν is measured has a rather close analog in classical mechanics. To the system described in the previous paragraph, we apply an additional magnetic field H_1, which is much smaller than H_0 and perpendicular to it in direction. If we consider the initial condition such that H_1 is perpendicular to both the angular momentum and H_0, the additional precession caused by H_1 will be such as to increase or decrease the angle between the angular momentum, J, and H_0, depending on the relative directions. If H_1 rotates with the frequency ν this effect is cumulative and the change in angle can be made large. It is apparent that if the frequency of revolution, f, of H_1 about H_0 is markedly different from ν, the net effect will be small. . . . The smaller the ratio H_1/H_0 the sharper this effect will be in its dependence on the exact agreement between the frequency of precession, ν, and the frequency f.

"Any method which enables one to detect this change in orientation of the angular momentum with respect to H_0 can therefore utilize this process to measure the precession frequency and therefore the magnetic moment. . . .

"The precise form of the initial conditions previously described is not important and we may consider H_1 initially at any angle ϕ with the plane determined by H_0 and J but still perpendicular to H_0. . . .

"In practice it is frequently more convenient to use an oscillating field H_1 rather than a rotating field. Although the situation is not quite as clear as for the rotating field, it is reasonable to expect that the effects will be similar if the oscillating field is sufficiently small. A simple calculation shows that no change in the magnitude of the projection of J on H_0 will occur unless the frequency of oscillation is close to the frequency of precession." Since the method depends on this matching of frequency, it is referred to as a *resonance method*.

"The arrangement used in our experiment is shown precisely in Fig. [6-]1. A stream of molecules coming from the source, O, in a high vacuum apparatus is defined by a collimating slit, S, and detected by some suitable device at D. The magnets, A and B, produce inhomogeneous magnetic fields, the gradients of which,

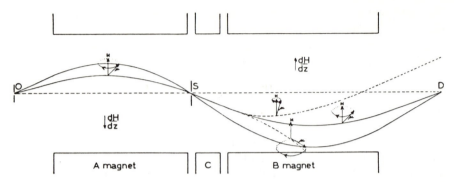

Fig. 6-1. Schematic arrangement of the apparatus for the resonance molecular-beam measurement of nuclear magnetic moments. [*Phys. Rev.* **55** (1939), p. 527, Fig. 1.]

$d|H|/dz$, are indicated by arrows. When these magnets are turned on, molecules having magnetic moment will be deflected in the direction of the gradient if the projection of the moment, μ_z, along the field is positive, and in the opposite direction if μ_z is negative. A molecule starting from O along the direction OS will be deflected in the z direction by the inhomogeneous A field and will not pass through the collimating slit unless its projected amount is very small or it is moving with very high speed. In general, for a molecule having any moment, μ_z, and any energy, $\frac{1}{2}mv^2$, it is possible to find an initial direction for the velocity of the molecule at the source such that the molecule will pass through the collimating slit. This is indicated by the solid lines in the diagram. If d_A denotes the deflection at the detector from the line OSD suffered by the molecule due to the field A alone, it may be expressed by:

$$d_A = (\mu_z/2mv^2)\,(d|H|dz)_A\,G_A\text{.''}$$

Here G_A is a factor depending only on the geometry of the apparatus. "The deflection in the B field will be in a direction opposite to that in the A field and is given by:

$$d_B = (\mu_z/2mv^2)\,(d|H|/dz)_B\,G_B.$$

The factors, $(d|H|/dz)_A G_A$ and $(d|H|/dz)_B G_B$, ... can be adjusted to have the same value. Thus if a molecule has the same μ_z in both deflecting fields it will be brought back to the detector by the B field. ...

"Magnet C produces the homogeneous field H_0. In addition, there is a device, not pictured in Fig. [6-]1, which produces an oscillating field perpendicular to H_0. If the reorientation which we have described takes place in this region the conditions for deflecting the molecules back to D by means of the B magnet no longer obtain. The molecule will follow one dotted line or the other depending on whether μ_z has become more positive or has changed sign." This is indicated by the small gyroscopes drawn on one set of the paths. "In fact, if any change in orientation occurs, the molecule will miss the detector and cause a diminution in its reading. We thus have a means of knowing when the reorientation effect occurs."

The earlier discussion in classical terms, while it gives a qualitative feel for how the method operates, is inadequate for quantitative considerations. Classically, detailed analysis yields an oscillation of the angular momentum vector around its unperturbed motion when the applied frequency f is different from ν, and a linear change in the angle of the precession cone when $f = \nu$. In quantum mechanical terms, the situation is that the component of the angular momentum in the direction of the external field can have only the discrete set of values $J, J - 1, J - 2, \ldots -J + 1, -J$, and that the oscillating field induces transitions from one of these values to another. The quantity of importance is the probability of such a transition as a function of time. The explicit form of this function will not be given here. It is a complicated polynomial, of order $4J$, in sines and cosines of an angle α which itself is a function of t of the form $A\sin\Omega t$, with A and Ω known functions of the ratio q of f to ν [more properly, to $\nu' = \mu(H_0^2 + H_1^2)^{1/2}/Jh$] and of the ratio of H_1 to H_0. The resonance occurs when $f = \nu'$; for that condition, $A \cong 1 - H_1^2/4H_0^2$, and $\Omega = \pi\nu'H_1/2H_0$.

"The orders of magnitude involved can be seen from a simple example: consider a system with spin $\frac{1}{2}$ and a moment of 1 nuclear magneton [$\cong 0.5 \times 10^{-23}$ cgs units] in a field of 1000 gauss and an oscillating field of 10 gauss amplitude. We assume that the system is moving at a speed of 10^5 cm per second which is of the order of thermal velocities, and set $t = l/v = 10^{-5}l$. The resonance frequency is[10]

$$\frac{\mu H}{hJ} = \frac{(0.5 \times 10^{-23})\,(10^3)}{(6.55 \times 10^{-27})\,(l/2)} \sim 1.5 \times 10^6 \text{ cycles per sec.,}$$

which fortunately is in a very convenient range of radiofrequencies. To make the $\sin^2$ terms a maximum at $q = l$ we set[11]

$$\pi \times 10^{-5}l \times 1.5 \times 10^6 \times 0.5 \times 10^{-2} = \pi/2.$$

Solving for l, we obtain $l = 6.6$ cm, which is a very convenient length for the oscillating field."

Rabi et al. go on to point out that for the study of nuclear moments, it is preferable to use atoms or molecules with zero electronic angular momentum, since otherwise the desired effects would be overwhelmed by effects due to the electronic moment. On the other hand, in atoms or molecules of zero electronic moment, the nuclear moment is coupled weakly enough to other influences, compared with its coupling to the external field, that it may be regarded as essentially free.

The diagram of the actual apparatus is given in Fig. 6-2. It "is contained in a long brass-walled tube divided into three distinct chambers, each with its own high vacuum pumping system. The source chamber contains the oven which is mounted on tungsten pegs. By means of a screw the mount may be moved, under vacuum, in a direction perpendicular to the beam axis. . . . The interchamber . . . provides . . . vacuum isolation of the receiving chamber from the gassing of the heated oven, by means of a narrow slit on each end of the chamber. These slits may be moved under vacuum in a manner similar to the oven mount. The receiving chamber contains most of the essential parts of the apparatus: the two deflecting magnets, *A* and *B*, the magnet, *C*, which produces the constant field, the radio-frequency oscillating field, *R*, the collimating slit, *S*, and the 1-mil tungsten filament detector, *D*.

"The *A* and *B* fields are electromagnets of the type . . . [used in an earlier method and shown in Fig. 6-3]. The gap is bounded by two cylindrical surfaces. . . . The gap width in the plane of symmetry, defined by the axes of the two cylindrical surfaces, is 1.0 mm. . . . A current of 300 amperes in the windings yields a field of over 12,000 gauss and a gradient of about 100,000 gauss/cm in the gap.

"The *C* magnet . . . is of conventional design. It is wound with 12 turns of 3/16″ square copper rod. . . . The pole faces, separated by a gap of 1/4″, are 10 cm long. . . . A field of about 23 gauss is realized in the gap per ampere of current in the exciting coils. . . .

"The oscillating field, *R*, consists of two 1/8″ copper tubes, 4 cm long, carrying current in opposite directions. These tubes are

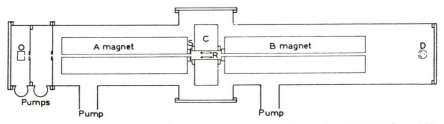

Fig. 6-2. Diagram of the actual apparatus. [*Phys. Rev.* **55** (1939), p. 529, Fig. 2.]

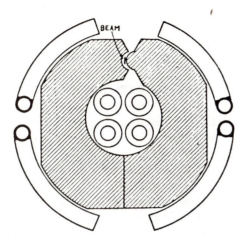

Fig. 6-3. Cross section of one of the deflecting magnets and its windings. [*Phys. Rev.* **53** (1938), p. 386, Fig. 2.]

flattened to permit their insertion between the pole faces of the C magnet when a space of about 1 mm between the tubes is left for the passage of the beam. The plane defined by the centers of these tubes is horizontal and is adjusted to be closely the same as the planes of symmetry of the A and B magnets. These tubes are supported by heavy copper tubing through which electrical and water connections may be made outside the apparatus.

"The magnetic field, H_1, produced by a current in the tubes is about 2 gauss/amp. and is approximately vertical and therefore at right angles to the field H_0 produced by the C magnet. The high frequency currents in the tubes are obtained by coupling a loop in series with them to the tank coil of a conventional . . . oscillator. . . . The frequencies used for these experiments range from 0.6 to 8 megacycles. The currents producing the oscillating field may be varied from 0 to 40 amperes."

The real secret of the success of the method was the painstaking adjustments of the apparatus. Space does not permit describing all of these in detail; a sampling will have to give the flavor.

"In mounting the magnets in the apparatus, care must be taken to avoid regions of weak, rapidly changing fields between magnets. Such regions cause transitions between quantum states . . . and prevent good refocusing of the beam by the B field. . . . The magnets are placed as close to each other as the windings will permit. Moreover . . . slabs of iron [are mounted] as extensions on both ends of the C magnet and on the ends of the A and B magnets facing the C magnet. This arrangement ensures a fairly strong field along the entire path of the molecular beam where changes in the over-all magnetic moment of the molecule affects its position at D. . . .

"A preliminary line-up of magnets, slits and detector is made by optical means while the apparatus is assembled. If this line-up is sufficiently good, a beam may be sent through the apparatus and a more precise line-up made. . . .

". . . Since it is impossible to sight through the gaps with a telescope, the preliminary optical line-up is made" by sighting on specially placed extensions. In this way the directions of the four fields—A, B, C, and rf—are adjusted, and the gaps are positioned well enough to permit a beam to pass.

"In the present apparatus the B magnet is permanently fixed inside the vacuum chamber and all other line-up operations are made with respect to it. By suitable movements of the oven, the collimating slit, and the detector, the beam is shifted . . . into a position at which one would like to have the plane of the edges of the A field. This field is then moved . . . until its fiduciary edges cut the beam. . . . The beam is then translated" to the desired final position.

The discussion of the principles of operation indicated that the perturbing frequency was the variable. However, "because the amplifier is not completely shielded from the oscillator and because the steady deflection of the galvanometer associated with the amplifier due to the oscillator is a function of the frequency, observations are made of the beam intensity as a function of the magnetic field, H_0, when the frequency is held fixed." An example of one such curve, that for ^{7}Li, is shown in Fig. 6-4.

"Since the value of the magnetic moment of any nucleus is calculated from an observed magnetic field and an observed frequency it is essential that these quantities be known to a high degree of precision. The frequency of the oscillating magnetic field is determined

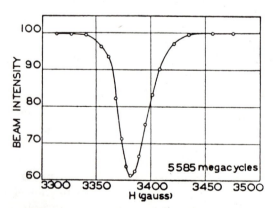

Fig. 6-4. Resonance curve for ^{7}Li, observed in LiCl. [*Phys. Rev.* **55** (1939), p. 532, Fig. 4.]

to better than 0.03 percent by measuring . . . with a . . . heterodyne frequency meter. . . .

"A calibration of the magnetic field of the homogeneous C magnet in terms of the current through the exciting coils was made in the usual way by measuring the ballistic deflection of a galvanometer[12] when a flip coil was pulled from the magnetic field. The galvanometer was calibrated by the use of a 50-millihenry mutual inductance, good to 1/2 percent. Several flip coils were constructed . . . on carefully measured brass spools. Errors in the magnetic field due to uncertainties in flip coil areas are probably not greater than 0.2 percent. . . .

"It is important that the magnetic field always return to the same value for a given magnetizing current. It was found that when a definite, reproducible procedure was used for demagnetizing the homogeneous field and for bringing it up to any state of magnetization, this condition was fulfilled to better than 0.1 percent.

"A considerable variation in the value of the mutual inductance was observed. . . . The absolute value of the magnetic field is indeterminate to about 0.5 percent. . . .

"The first nuclei to be studied by this method were $_3Li^6$, $_3Li^7$ and $_9[F]^{19}$ in the LiCl, LiF, and NaF and Li_2 molecules. . . . For each nucleus the f/H values corresponding to the resonance minima are constant to a very high degree for wide variations of frequency. This shows that we are dealing with a change of nuclear orientation and not with some molecular transition. . . .

"The nuclear g is obtained from the observed f/H values by use of the formula

$$g = \frac{4\pi}{e/Mc} \cdot \frac{f}{H} = 1.3122 \times 10^{-3} \frac{f}{H} ,$$

which follows immediately from Eq. ([6-]1) [p. 88] if the magnetic moment μ is measured in units of $eh/4\pi Mc$, the nuclear magneton, and $f = \nu$. . . . The nuclear g's, the spins and the magnetic moments are listed in Table [6-I]."

Table 6-1. Nuclear g's and magnetic moments determined in the first molecular-beam resonance experiment. [*Phys. Rev.* **55** (1939), p. 534, Table II.]

Nucleus	g	Spin	Moment
$_3Li^6$	0.820	1	0.820
$_3Li^7$	2.167	3/2	3.250
$_9F^{19}$	5.243	1/2	2.622

Rabi and his coworkers had clearly fulfilled their aim of providing a method whereby nuclear moments could be determined to high precision. Not long afterwards, a modification of the method was used to determine the magnetic moment of the neutron. Only then was the precision enough to establish that the magnetic moment of the deuteron is not simply equal to the sum of the magnetic moments of the proton and the neutron, but that there are contributions from dynamical effects.

The molecular-beam resonance method has since been supplemented, and exceeded in precision, by other techniques. But it remains an important approach; and in the newer methods, the resonance concept continues to have a major role.

Notes

1. See *Crucial Experiments*, chap. 5.

2. The "fine structure" pertains to the separation within multiplets, as between the yellow "D" lines of sodium; the "hyperfine structure" refers to still smaller splittings.

3. This is what would be expected from order-of-magnitude estimates: The force on a charged particle moving with speed v in a magnetic field B is approximately $q(v/c)B$, and the magnetic field at a distance R from a magnetic dipole of moment μ is of order μ/R^3, so that the force is $q\mu v/cR^3$. Nuclear magnetic moments should be of order $e\hbar/m_N c$, where m_N is the mass of a nuclear constituent particle (proton, say), about 2000 times the mass of the electron. Thus the magnetic force is $qe\hbar v/m_N c^2 R^3$, and the electric force is qe/R^2, so that the magnetic force is weaker by the ratio $\hbar v/m_N c^2 R$. The distance R is at least equal to the nuclear radius, which is of the same size as the "classical" radius of the electron, $e^2/m_e c^2$; thus the ratio is roughly $\hbar v m_e c^2/e^2 m_N c^2$. The combination $e^2/\hbar c$ is a well-known dimensionless constant, approximately $1/137$, so the ratio of magnetic to electric force is roughly $137(v/c)(m_e/m_N) = \frac{1}{10}v/c$, clearly a small number.

4. Goudsmit comments, "The magnetic moment was not mentioned explicitly since it was implied, just as is the case with the magnetic moment of the atom."

5. They ascribe the failure primarily to the likelihood that only some mercury isotopes—that of odd mass number, which constitute only about 30% of the total—can be expected to have magnetic moments.

6. See *Crucial Experiments*, chap. 8.

7. In addition to the conceptual developments, Rabi was apparently responsible for a significant experimental modification: replacing the strong but spatially constricted magnetic fields of permanent magnets with weaker but more extensive fields produced by currents in long wires.

8. All angular momenta are expressed in units of $\hbar$.

9. According to quantum mechanics, the angular momentum and with it the magnetic moment make a nonzero angle θ with the direction of the magnetic field. There is therefore a torque $\mu H_0 \sin\theta$ in a direction perpendicular to both H_0 and J; this torque causes the precession.

10. Some misprints in the original of the following equation have been corrected.

11. This is just the condition $\Omega t = \pi/2$.

12. See note 7 in Chap. 4, p. 55.

Bibliography

The detailed report of the first experiment is I. I. Rabi, S. Millman, P. Kusch, and J. R. Zacharias, *The Physical Review* 55, 526 (1939). This and several other papers on molecular beams are reprinted in *Molecular Beams: Selected Reprints. Vol. I, Experiments with Molecular Beams* and *Molecular Beams: Selected Reprints. Vol. II, Atomic & Molecular Beam Spectroscopy* (American Institute of Physics, New York, 1965). Both volumes also contain an annotated bibliography of further works on molecular beams.

Chapter 7

Fine Structure in the Spectrum of Hydrogen

The conventional story of the development of the theory of the hydrogen atom is slightly misleading. The great triumph of Bohr's theory was its confirmation of the Balmer formula; yet only a year after Balmer's work, and thus over 25 years before Bohr, the spectrum was found to be more complex than Balmer's formula accounted for. A. A. Michelson and E. W. Morley discovered that the first line of the series, H_{α}, is a doublet, with a separation in wave number of 0.36 cm^{-1}.

Of course, it was not surprising that Bohr's theory failed to do everything perfectly; the wonder, rather, was that it did so much so well. Modifications and improvements were not long in coming. In particular, Arnold Sommerfeld in 1916 took account of the effects of relativity on the motion of the electron and found that for $n = 2$ (the final state in the transition that gives the H_{α} line), there would indeed be two possible motions, whose energies differed by

$$\Delta W_2 = \tfrac{1}{16} \alpha^2 hcR[1 + O(\alpha^2)], \tag{7-1}$$

where $h \equiv 2\pi\hbar$ is Planck's constant, c is the speed of light, $R = 109,737.3$ cm^{-1} is the Rydberg wave number, $\alpha = e^2/\hbar c \cong 1/137$ is a dimensionless number called the *fine structure constant* (e is the charge of the electron), and the notation $O(\alpha^2)$ means terms of relative order of magnitude α^2 or smaller. There would also be a threefold splitting of the level with $n = 3$, but this would be too small to be detected. Since Eq. (7-1) gave the right value for the doublet separation, the situation was satisfactory for the time being.

Oddly enough, the introduction of the new quantum theory made matters worse at first. The Bohr formula for the main level structure was reproduced, of course; but attempts to include relativistic effects gave a value for ΔW_2 that was too large by a factor 8/3. This state of affairs was relieved somewhat by the introduction of the concept of electron spin and the associated magnetic moment. Calculations including these effects now gave the results shown in Fig. 7-1. For $n = 2$ there were three substates but only two energy values, as the energy was found to depend on the total angular momentum J but not on the "orbital" angular momentum L.[1] Similarly, for $n = 3$ there were five states with three energy values. Moreover, the calculation reproduced Eq. (7-1) for ΔW_2, as derived from a general expression for $W_{n,J}$ that also gave the proper separation of the levels with $n = 3$.

However, the calculation for $L = 0$ was ambiguous. If the expression for the spin effects was evaluated by first inserting the relations $S = 1/2$ and $J = L + 1/2$ and reducing, there was no difficulty; but if instead the values $J = 1/2$, $L = 0$, and $S = 1/2$ were substituted directly, the ratio 0/0 appeared. Accordingly, it was reassuring when in 1928 P. A. M. Dirac developed a relativistic theory of the electron which automatically implied the existence of the spin and magnetic moment, and which gave an expression for $W_{n,J}$ that unambiguously reduced to the earlier expression to lowest order in α and thus, again, reproduced Eq. (7-1). However, the Dirac theory had some disturbing features of its own, and it was therefore still important to try to check experimentally the predictions regarding the hydrogenic spectrum.

The obvious approach would be to try to observe the fine structure of the H_α line by direct optical spectroscopy. This method is limited in accuracy by the fact that most optical sources operate at effective temperatures high enough to give the atoms a substantial thermal motion; the resulting Doppler effect gives the lines a width that is comparable with the separation expected. Consequently, different workers obtained conflicting results. Some found discrepancies with the theoretical predictions of a sort that could be explained, according to a suggestion made by Simon Pasternack, if the $2^2S_{1/2}$ level[2] were higher than the $2^2P_{1/2}$ by about 0.033 cm^{-1}; others found no discrepancy. Since Pasternack's suggestion was purely empirical, the question remained open.

The final solution was achieved by a method proposed in 1928 by W. Grotrian.[3] He noted that the selection rules[4] for atomic transitions gave no restrictions on n, so that transitions within the $n = 2$

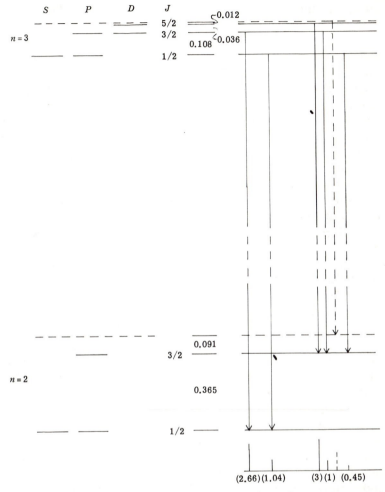

Fig. 7-1. Structure of the $n = 2$ and $n = 3$ levels of hydrogen. The structure within the two groups is shown to the same scale; the actual separation between the groups is about 100,000 times the total spread of the upper group. The numbers are the intervals in reciprocal centimeters (1 cm^{-1} is equivalent to 1.234×10^{-4} eV). The dashed levels show the positions as computed from the Bohr formula; the solid lines are the values obtained when spin and relativistic effects are included. On the left, the various values of L are separated. The right-hand part is a coalesced version showing the optical transitions, with vertical bars below on a wave-number scale showing ideal relative amplitudes. The first and fourth most intense components, which are not resolved optically, constitute the "main line" of the "doublet"; the second and third, which are barely resolvable, together make up the other part of the doublet. The total spread in wave number is 0.473 cm^{-1}, while the average wave number is about 15,233 cm^{-1}.

group should be possible. The appropriate frequencies would be of the order of 10 gigahertz (1 gigahertz = 10^9 cycles per second), corresponding to wavelengths of about 3 cm. There were even some attempts in the early 1930s to produce the suggested transitions, but radiofrequency techniques at such frequencies were not yet adequate. During World War II, however, substantial advances were made in microwave methods; and shortly after the war, Willis Lamb and Robert Retherford, at Columbia University, devised and carried out a means of making this direct measurement. The result, announced briefly in *The Physical Review* in 1947 and eventually presented in detail in a series of Articles in that journal, was that "contrary to [then current] theory but in essential agreement with Pasternack's hypothesis, the $2^2 S_{1/2}$ state is higher than the $2^2 P_{1/2}$ by about 1000 Mc/sec. (0.033 cm^{-1}) or about 9 percent of the spin relativity doublet separation." Their experiment, which will be described in this chapter, was an important stimulus to the development in the succeeding two years of a successful relativistic, quantized theory of the interaction of the electron with the electromagnetic field; and this work, together with subsequent related work of Lamb and coworkers, brought Lamb a Nobel prize for physics in 1955.

The experiment was far from simple. The series of detailed papers, especially the first, provides (over the protests of the editor of *The Physical Review* at that time) an exposition of the thinking and planning that went into it, of a sort that is practically unmatched in modern literature and that gives valuable insight into the way research is actually done. Lamb and Retherford "first inquired whether microwave techniques ... would now permit a successful and clear-cut determination of the hydrogen fine structure by absorption of the appropriate radiation in a ... discharge. . . . [T]he indication was rather discouraging." Estimates of the absorption cross section and the number of excited atoms that might be present indicated that the amount of absorption due to atomic transitions would be insignificant, whereas there would be a large background of absorption by the free electrons in the discharge. Meanwhile, Lamb and Retherford hit upon a different approach.

"There are two main methods in radiofrequency spectroscopy. In one, some matter absorbs or otherwise affects the radiation passing through it. In the other, the radiation causes observable changes to take place in the matter. The former method is used in the conventional form of microwave spectroscopy in which the radiation is transmitted through a long path of absorbing gas. The latter method is used in the molecular beam radiofrequency resonance method. . . .[5]

Since the first method did not seem to be too promising for the study of the hydrogen fine structure, we turned to an examination of the possibilities in the second method."

The key to success lay in a marked difference in the properties of the states involved. "In the case of atomic hydrogen, the $2P$ states decay to the $1S$ [ground] state with the emission of a photon of wavelength 1216A in 1.595×10^{-9} sec., and the atom could move only about 1.3×10^{-3} cm in that time, assuming a speed of 8×10^5 cm/sec. On the other hand, the possibility exists that the $2^2S_{1/2}$ state would be sufficiently metastable so that a beam of particles in this state could be used." The metastability would come about because this state cannot undergo ordinary radiative decay to the ground state, there being no way to supply the one unit of angular momentum that would have to be carried by the photon. "If a transition were then induced by radiofrequency or otherwise to a $2P$ state, a decay to $1^2S_{1/2}$ would take place so quickly that the number of excited atoms in the beam would be reduced. If one could then find a detector which responded selectively to the excited hydrogen atoms, one would have the possibility of measuring the energy difference between the metastable $2^2S_{1/2}$ state and the various $2P$ states. . . .

"For this program to be carried out it was necessary to solve two problems: (1) production of a beam of atoms in the $2^2S_{1/2}$ state and (2) detection of such atoms. The program clearly depended on a knowledge of the properties of metastable hydrogen atoms." There was not much of this knowledge to be had.

The first question, in fact, was the extent to which metastability actually showed up in realistic situations. For a truly isolated atom there was no doubt. The most probable fate is decay by emission of two photons; this gives a lifetime of the order of 1/10 second, more than adequate for the purpose. However, if the S and P states are actually degenerate, then the presence of an electric field causes the actual states to be neither pure S nor pure P, but a mixture of the two,[6] with a lifetime much smaller than that of the pure S state.

"For an effective electric field of 10 volts/cm, . . . the life of the $2^2S_{1/2}$ state would be about five times the life τ_p [of the $2^2P_{1/2}$ state, 1.6 nanoseconds]. Over a large range of fields, the life varies inversely as the square of the field strength. . . .

"It was clear to us from the foregoing that the $2^2S_{1/2}$ state would be markedly metastable only if the perturbing fields could be sufficiently reduced. For a beam length of 6 cm and a speed of 8×10^5 cm/sec., a life of the order 0.75×10^{-5} sec. would be necessary for 37 percent of the atoms to survive. According to . . . calculations, this would require a perturbing field of 1/3 volt/cm or less. It would not be easy to keep the electrons and ions formed in the excitation process away from the detector with such fields. . . .

"In actuality, of course, we now know that the $2^2S_{1/2}$ and $2^2P_{1/2}$ levels are not degenerate. This adds considerably to the stability of $2^2S_{1/2}$ against Stark effect quenching. . . . At the time, however, we did not take the possibility of a natural removal of the degeneracy by [any substantial] amount very seriously, and planned to increase the stability of $2^2S_{1/2}$ by the application of a magnetic field to give a large Zeeman splitting[7] of the states." Since the splitting is different for S than it is for P, this artificially removes the degeneracy. "The presence of a magnetic field at right angles to the atomic beam would also serve to keep charged particles away from the detector"—thus also removing the need for an undesirable electric field.

The next problem was that of producing a beam of atoms in the metastable state. "A number of possible methods . . . were considered. The simplest source would be a hydrogen discharge tube with a small opening into a vacuum for the emergence of a beam. In the discharge, one would have a mixture of molecular and atomic hydrogen, electrons and ions, and a small proportion of excited atoms, together with a high intensity of Lyman and Balmer radiation. . . . The question would be whether any appreciable number of [$2^2S_{1/2}$] could escape through the hole. . . . It would also be possible for some of the normal atoms to be excited optically . . . after leaving the discharge, and . . . decay to $2^2S_{1/2}$. The numerical estimate of the yield . . . was discouraging. Furthermore, there would be a very high background intensity of ultraviolet radiation at the detector. This could cause trouble since most of the possible detectors of metastable atoms are also photosensitive. . . .

"The second method which we considered was the bombardment of molecular hydrogen with electrons in a field-free region. . . . [F]ast metastable atoms should be produced at . . . bombarding energies [above about 25 eV].

"One difficulty with this method of production is that the atomic fragments H + H* move off in directions oriented at random with respect to the electron beam, and hence the detector can conveniently intercept only a small fraction of them. The background effects due to the ultraviolet photons which are produced in molecular hydrogen beginning at 11.5 volts are likely to be troublesome. . . .

"The third method, which was finally adopted, required the separate production of a beam of atomic hydrogen atoms in the ground state and the subsequent bombardment of the beam by electrons with an energy somewhat over the threshold of 10.2 volts for excitation to $2^2S_{1/2}$. . . .

"A number of methods are available for the production of a beam of atomic hydrogen atoms: (1) Wood's tube, (2) microwave discharge, (3) thermal dissociation in a tungsten furnace." The last was chosen, for "no very compelling reason." Estimates after the fact indicated oven pressures of 10^{-3} atmosphere; an oven temperature of 2500 K then gave about 65% dissociation and a most probable speed in the original beam of 8×10^5 cm/sec.

The excitation of the atoms by electron bombardment was not so well assured. "The cross sections for various electron excitation processes in hydrogen have been calculated by Bethe according to the Born approximation, with neglect of electron exchange. Although this approximation is not expected to be very good for electron energies near the threshold, it was the only one available for our numerical estimates." The maximum cross section for excitation of 2^2S in this approximation is 2.2×10^{-17} cm^2 and occurs at 14.8 volts, but the yield "does not fall below half the maximum down to 11 volts.

"Besides the errors inherent in the Born approximation, the neglect of electron exchange may be particularly serious . . . and one may hope to obtain a much larger cross section with a maximum nearer the threshold. . . . Since no reliable calculations . . . seem to exist . . ., we used the lower threshold $\sigma = 10^{-17}$ cm^2 hoping to be on the safe side."

The process of bombardment itself caused some complications.

"One must . . . choose between bombardment at right angles to the beam, along it, or against it. The last two methods would not destroy the unidirectionality of the beam of atoms if the bombardment were at a voltage just above threshold. Such a choice would be natural for other atoms, but not for metastable hydrogen atoms. In this case, one would be forced to have the magnetic field parallel to the electron beam.[8] If the electrons were sent against the atomic beam, the metastable atoms would subsequently pass through an electric field [in the electron source] which would quench them. If the electrons were sent along the beam, they could not easily be kept from the detector. . .

"The first choice necessarily leads to a transverse recoil, and what is worse, an indefinite one, so that an originally well-collimated beam is diffused by the bombardment. This means that the use of slits of the order of 0.001 in. width as in the usual atomic beam work is ruled out." But the other alternatives appeared unworkable, and so these consequences were accepted.

The next question was that of detection. Again, "several possible methods . . . were considered." One possibility would have been the detection of the radiation given off in the decay of the $2^2P_{1/2}$ state (the Lyman α line, 1216 Å). This was not attempted with hydrogen, although the method was later used in a similar experiment on ionized helium. Other possibilities were that a metastable atom striking a suitable metal surface would either eject an electron or be ejected itself as an ion. Each of these processes was known to occur for some metastable species. The necessary condition for the former is that the excitation potential of the metastable state (10.2 V for $2^2S_{1/2}$) be higher than the work function of the metal; for the latter, that the ionization potential (3.4 V) be less than the work function. Both conditions were likely to be satisfied for "most surfaces which can be maintained under the conditions of the experiment." "According to rough calculations . . ., metastable hydrogen atoms of thermal energy moving in toward a metal surface should, on the average, cause electron ejection before the atom reaches a distance of 2A. Since half of the electrons emitted probably go into the metal, one might expect an efficiency of the order of 50 percent.

". . . [W]e found that electron ejection did occur, but probably not with the expected high efficiency. We have looked unsuccessfully for positive ion emission." All this, however, was after the fact. Almost nothing was known in advance about the operation of these methods with metastable hydrogen.

Nevertheless, it was necessary "to make a rough estimate of the yield of metastable atoms and the resulting electron currents to be expected under typical experimental conditions." The detector signal S depends on several factors: the rate of escape of atoms from the oven in such a range of directions as to reach the detector, the fraction of these that are excited to $2^2S_{1/2}$, the fraction δ of metastables that recoil in the proper direction to reach the detector, the fraction μ that survive their passage through any quenching electric fields, and the efficiency η of electron ejection from the detector surface by a metastable hydrogen atom. The expression for the signal *without* application of microwaves is

$$S = n_0 a X A I \sigma \delta \mu \eta /(4\pi R^2 h) \text{ electrons/sec,}$$

where, in addition to the symbols defined above, the following are used: n_0 is the number of atoms and molecules per unit volume in the tungsten oven, a is the area of the opening in the wall of the oven, X is the fractional dissociation, A is the area of the detector, I is the bombarding current in electrons per second, σ is the cross section for exciting the atoms from the ground state to the metastable state, R is the distance of the detector from the oven slit, and h is the height of the electron beam.

"Unfortunately, many of the factors needed for the estimate of the yield were not known, and a certain degree of optimism was required in the assumption of relatively favorable values for them in order to predict a usable signal." In particular, "the value for σ was based on very inadequate theory. The value $\mu = 1/2$ for the metastable survival factor was reasonable . . . provided quenching effects were not present. . . . The detector efficiency $\eta = 0.50$ was taken most optimistically. . . .

"With the above assumptions, the signal is

$$S = 3.26 \times 10^7 \text{ electrons/sec.}$$
$$= 5.2 \times 10^{-12} \text{ amp.}$$

This is a current which is about 5×10^4 times the least current $(10^{-16}$ amp.) which can be detected conveniently with an FP54 electrometer circuit[9] and a sensitive galvanometer. Consequently, unless one or more of the estimates proved to be too optimistic, the signal should be large enough to detect, and even to use for accurate radiofrequency spectroscopy." As it turned out, the estimate "proved to be very optimistic"—but not by orders of magnitude.

The next consideration was the amount of rf power required. "The metastable atoms are to be subjected to radiofrequency waves somewhere between the source and detector. When the frequency is such that $\hbar\omega$ is equal or nearly equal to an energy difference between a Zeeman component of the $2^2S_{1/2}$ level and a Zeeman component of one of the $2P$ levels, the radiation may induce a transition to a non-metastable level and the detected signal will decrease. The amount of decrease will depend on the intensity of the radiation as well as its frequency, on the speed of the atoms, and the length of the radiofrequency region." The calculation then went as follows: If "the decay rate of state $2^2S_{1/2}$ to one of the $2P$ states due to r-f is $1/\tau_s$ . . .," then "[t]he fraction of atoms quenched by the r-f field is

$$\Phi = 1 - \exp(-l/v\tau_s),$$

where l is the length of the r-f region, and v is the speed of the atoms. For $l = 1$ cm and $v = 8 \times 10^5$ cm/sec., the beam would be 63 percent quenched for

$$\tau_s = l/v = 1.25 \times 10^{-6} \text{ sec."}$$

The decay rate $1/\tau_s$ could be computed from the quantum theory of radiation; for example, for the transition from $2^2S_{1/2}(m = 1/2)$ to $2^2P_{1/2}(m = -1/2)$, with radiation polarized perpendicular to the magnetic field, and for the resonance frequency, the result is

$$1/\tau_s = (8\pi e^2 S_0/c\hbar^2\gamma)(3a_0^2),$$

where S_0 is the incident intensity, $\gamma = 1/\tau_p$ is the decay rate for P states, and $a_0 = 0.53$ Å is the Bohr radius. To give τ_s the required value, then, "requires an energy flux density at resonance of $S_0 = 3.4$ mw/cm². Since this power density is easily obtained at any frequency up to 30,000 Mc/sec., we did not anticipate any difficulty with r-f power requirements."

The object of the whole experiment, it will be recalled, was to determine the resonance frequencies. "The most obvious way . . . would be to measure the beam intensity as a function of radiofrequency, keeping the rf power constant. Unfortunately, this is next to impossible to do." Microwave oscillator outputs and transmission-line characteristics vary with frequency; even devices that might be used to monitor the rf power actually received at the quenching region are frequency sensitive.

"Since a magnetic field was believed to be necessary anyway, we decided to take advantage of it to overcome the above difficulties. The frequency (and power level) of the oscillator is kept constant and the atomic energy levels are moved through resonance by varying the magnetic field." The change in resonance frequency as the magnetic field is changed results from the change in the Zeeman splitting; it implies the need to measure the field accurately, a standard operation, and to know how the splitting varies. There is no need to discuss here the detailed theory of the Zeeman effect. For the hydrogen atom, all the calculations are quite straightforward. The result is a pattern of levels shown in Fig. 7-2. The corresponding transition frequencies, with allowance made for the selection rules $\Delta m = \pm1,0$, $\Delta J = \pm1,0$, are shown in Fig. 7-3, and equations relating the frequencies to the field are readily obtained and simple in form.

Two other effects enter into the situation. One is that with the shift of the $2^2S_{1/2}$ level that actually was found to exist, the pattern shown in Fig. 7-2 is modified by an upward shift of lines α and β, a distance of about 0.137 in units of y. As a result, levels β and e coincide—cross—at a field of about 540 gauss. Any electric field will then produce a Stark mixing of these two states and thereby destroy the metastability of the β state. Since the particles are moving perpendicular to a magnetic field, there is indeed an electric field $\mathcal{E} = (v/c)H$; for $H = 540$ gauss and $v = 8 \times 10^5$ cm/sec, $\mathcal{E} = 4.3$ volts/cm.

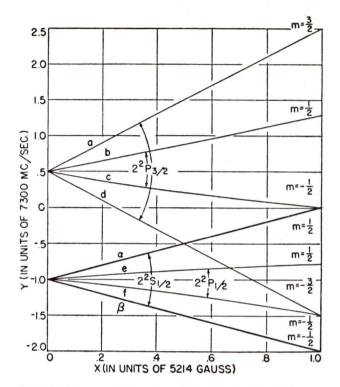

Fig. 7-2. Zeeman splitting of $n = 2$ levels in hydrogen. The vertical scale gives the energy in units of $\frac{2}{3}(E_+ - E_-)$ (the frequency equivalent of this unit is 7300 MHz), where E_+ and E_- are, respectively, the energies in zero magnetic field of the upper ($2^2P_{3/2}$) and lower ($2^2P_{1/2}$, $2^2S_{1/2}$) states; the zero point is the "center of gravity" of these, $\frac{1}{3}(2E_+ + E_-)$. The horizontal scale gives the magnetic field in units of $2(E_+ - E_-)/3\mu_0 = 5214$ gauss, with μ_0 denoting the Bohr magneton, $e\hbar/mc$. [*Phys. Rev.* **79** (1950), p. 559, Fig. 12.]

This gives a lifetime of 4.3×10^{-8} sec for state β at the critical field value. Moreover, "the state of affairs described above is not confined to the critical field of $H = 540$ gauss for which the levels β and e cross, but extends over quite a range of field strengths. With the apparatus to be described [here], the transitions βb, βc, βd, βe, βf, . . . were never observed.

"Motional Stark quenching also occurs about the magnetic field where α and c cross, or [with allowance for the shift] $H = 4700$ gauss. This region is much wider than for the βe crossing . . . and the beam of metastable atoms will be strongly quenched above 3000 gauss."

The other effect is that of interaction between the electron and the magnetic moment of the nucleus. This brings about a further

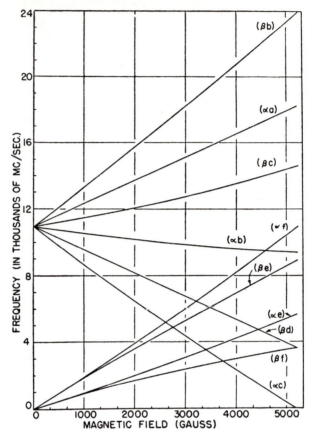

Fig. 7-3. Resonance frequencies expected within the n = 2 complex in hydrogen on the basis of Dirac theory, as functions of magnetic field. The pairs of letters designate the pairs of states involved in the transition, according to the notation shown in Fig. 7-2. [*Phys. Rev.* **79** (1950), p. 560, Fig. 13.]

splitting of the levels: Each of the levels shown in Fig. 7-2 is further split into two.[10] However, the splitting is small, corresponding at most to a frequency 88 MHz. The expected quenching resonances are not infinitely narrow but have a natural width because the quenching rate $1/\tau_s$ does not vanish for applied frequency ω equal to the resonance frequency ω_0 but is proportional to $1/[(\omega - \omega_0)^2 + (1/2\gamma)^2]$, where $\gamma = 1/\tau_p$. This produces a curve that is roughly bell-shaped, with a width of order $\gamma/2\pi = 99.8$ MHz. Thus the effect of the hyperfine interaction will be overpowered by the width of the line and will usually make it broader—especially the peak portion.

A preliminary apparatus was constructed "as simply as possible consistent with [the] aim" of establishing simultaneously the pro-

duction and detection of metastable atoms, since both of these processes were "theoretically possible, but experimentally unproven."
Source and detector "were enclosed in glass bulbs connected by a
glass tube 1/2 in. in diameter and 6 in. long. Magnetic fields could be
applied to the source, detector, and intervening region. . . . It was
thought that at a later time the connecting tube could be passed
through the broad side of an X band wave guide. . . .

"The . . . method . . . of direct excitation by electron bombardment of the molecule, was selected as probably the easiest way to
produce metastable hydrogen atoms. . . . [T]he electron ejection
method of detection was adopted. . . . Unfortunately, the bombardment of molecular hydrogen produced a variety of effects. . . [A]t
this time it was not possible to verify the presence of metastable
atoms."

It was established, however, that hydrogen atoms were being
formed and were reaching the detector. This was done by depositing
on the detector plate a layer of yellow molybdenum oxide "soot,"
which is reduced to a blue form by atomic hydrogen. The conclusions were "that the chosen excitation method was not as favorable
as anticipated" but that the overall program was promising enough to
warrant further effort. "It appeared that the two-step method of first
dissociating the molecule and then exciting the atoms by electron
bombardment would be less affected by [the] difficulty" of extrane-
ous processes that would overwhelm the effects due to metastables.

"Since it was clear that an extensive revision of the apparatus
was necessary, a primarily all metal one was built incorporating all
known improvements. Nevertheless, many modifications were made
before meaningful results were obtained.

"The general scheme of the apparatus is illustrated in Fig. [7-4]
which is a horizontal cross section through the axis of the coaxial
circular cylinders comprising the right- and left-hand chambers. The
rectangular space, J, is the cross section of an X-band rectangular
wave guide . . ., which passes through the apparatus vertically. The
magnetic field, neglecting fringing, is at right angles to both the wave
guide and the right and left chambers. The magnet pole pieces fit
into truncated conical indentations in the apparatus. . . . [T]he outer
shell of the apparatus is principally composed of . . . copper.

". . . Hydrogen atoms leave the hydrogen dissociator A and pass
to the interaction space D where some of them are excited to the $2s$
state by electron bombardment. After experiencing a small recoil,
they pass through the slits I and r-f region to the detector target, L.

". . . [T]he recoil angle was taken into account in the choice of
the relative positions of the hydrogen dissociator, the electron bombarder, and the slits I. . . .

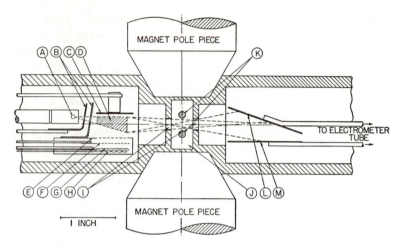

Fig. 7-4. Cross section of the apparatus. Letters indicate the following: *A*, oven of the hydrogen dissociator; *B*, shielding diaphragms; *C*, *E*, *F*, *G*, *H*, elements of the electron bombardment system; *D*, bombardment region; *I*, slits; *J*, wave guide; *K*, transmission lines; *L*, *M*, metastable detector. Further details are given in text. [*Phys. Rev.* **79** (1950), p. 564, Fig. 19.]

"The detection electrodes consist merely of two tungsten plates, *L* and *M*. . . [M]etastable atoms fall on *L* and eject electrons which are collected by *M* which is held 3 or 4 volts positive with respect to *L*. The electron current is measured by means of a standard FP54 electrometer circuit. . . . The response to metastable atoms is not very sensitive to collector voltage so long as it is 2 or 4 volts. . . .

"Atomic hydrogen was produced by thermally dissociating molecular hydrogen. . . . The details of the arrangement are shown in Fig. [7-5]. A thin-walled tungsten cylinder . . . was fabricated from . . . tungsten sheet as indicated, and a slot 0.008 × 0.060 in. was cut near the center and parallel to the cylinder axis. The cylinder was then forced into the holes provided in the molybdenum ends of the water-cooled leads. Molecular hydrogen was introduced through a tube inside one of the water ducts. The tungsten tube could then be heated in its central portion to a temperature sufficient to produce a satisfactory degree of dissociation. The atoms so produced stream out of the small slot which serves as the source. . . .

"Operating temperatures in the central portion of the cylinder were usually about 2500° K. . . .

"Alternating current was found to be satisfactory for heating the hydrogen dissociator. A current of 80 amp. and a voltage drop of 2 volts represent typical operating conditions. . . .

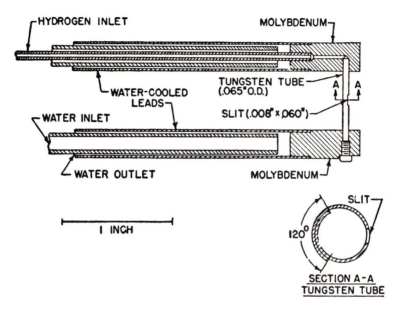

Fig. 7-5. Details of the hydrogen dissociator. [*Phys. Rev.* **79** (1950), p. 564, Fig. 20.]

"The electron bombarder has proven to be the most troublesome part of the apparatus, and much effort was expended in determining how to make a good one. . . . [I]t is quite possible to state the properties of an ideal electron gun for this experiment. . . . In practice, however, to obtain sufficient signal it is necessary to depart considerably from [the ideal] principles.

"The electron bombarder used consisted of a cathode, a control grid, an accelerator grid, and an anode. . . . The cathode, *G*, was of the oxide-coated, indirectly heated type. The cathode and control grid, *F*, were completely enclosed by the accelerator grid electrode, *E*, which had grid wires only on the side facing the anode. This served to shield the interaction space, *D*, where excitation occurred, from the electric fields between the accelerator grid and the cathode. The shields *B* prevented radiation and evaporated matter from the hydrogen dissociator from reaching the electron bombarder and detector. In the earlier models . . . the anode *C* was absent, its purpose being served by the outer shell. With such an arrangement no metastable atoms were ever detected with certainty." This was found to be the result of unexpected space-charge effects, which were greatly reduced by the smaller electrode spacing made possible by the introduction of the separate anode.

"It was found to be most advantageous to operate with an accelerator grid voltage of about 13.5 volts, and to bias the anode 3 volts positive with respect to the accelerator grid. The anode current was held at about 0.3 ma by adjusting the control grid voltage. . . . In general the operating conditions chosen represent a compromise between instability and signal strength. . . .

"The metastable atoms are subjected to r-f or d.c. fields in the wave guide region J. The atoms enter and leave the region through the slits, I. . . . The wires, K, were installed long before any r-f fields were applied. . . . [T]hey proved to be of great value in identifying metastable atoms" by Stark quenching in a dc field applied between them, leading to a reduction in the detector signal. "In view of the difficulty encountered in maintaining a good supply of metastable atoms this feature of the apparatus was found to be indispensible, and not a temporary measure as originally intended. Furthermore, the transitions to the $2^2P_{1/2}$ levels occur at frequencies of 1500 to 6000 Mc/sec. at which the wave guide is beyond cut-off.[11] For frequencies in that range the wires, K, were used as a two-wire transmission line. . . .

"Frequencies between 3000 and 10,000 Mc/sec. were obtained directly from . . . klystrons. Those frequencies near 12,000 Mc/sec. were obtained by doubling the frequency from a . . . klystron in a . . . crystal multiplier. Frequencies from 1500 to 2600 Mc/sec. were obtained from a lighthouse tube oscillator. The frequency stability resulting from the use of ordinary regulated power supplies was found to be sufficient. The frequency, or rather the wavelength, of the radiation was measured by means of coaxial or cavity wave meters.

"The magnetic field was produced by a small electromagnet. . . . Magnetic fields of 3000 gauss or more were readily obtained. . . .

"The field was calibrated by moving the detector electrodes and inserting a small search coil into the center of the r-f region through the slit, I. The usual standard demagnetization procedure and magnetization curve method were followed in obtaining the calibration curve of magnetic field strength H *versus* magnet current, and throughout the measurements. The calibration was made to an accuracy of about 0.5 percent. This was more than sufficient accuracy, for it was found that the field in the r-f interaction space was nonuniform by some 33 gauss out of a total of 1000. . . .

"The detector current consisted of the signal, or current of electrons ejected by metastable atoms, and the background or current of photo-electrons arising primarily from excitation of . . . background . . . hydrogen molecules, and to some extent from excitation of atomic states other than the $2^2 S_{1/2}$. At the electron energies used, the background was three or more times as large as the signal. . . .

"The signal was observed as the galvanometer deflection obtained when a d.c. field large enough to quench practically all the metastable atoms was applied to the wires K." Apparently, sometimes the alternative method of interrupting the rf was used. "This deflection . . . usually [corresponded to 0.3×10^{-13} to 0.75×10^{-13} amp] and on occasion as much as . . . 1.2×10^{-13} amp. This is smaller than the estimated value . . . by a factor of about 40. An undetermined part of this discrepancy could be attributed to quenching by stray electric fields." The remainder was presumably due to a substantial overestimate of the detector efficiency.

The preceding paragraph was extracted from the first paper of the detailed series. The original report is much more condensed: "We have observed an electrometer current of the order of 10^{-14} ampere which must be ascribed to metastable hydrogen atoms. The strong quenching effect of static electric fields has been observed, and the voltage gradient necessary for this has a reasonable dependence on magnetic field.

"We have also observed the decrease in the beam of metastable atoms caused by microwaves in the wave-length range 2.4 to 18.5 cm in various magnetic fields. In the measurements, the frequency of the r-f is fixed, and the change in the galvanometer current due to interruption of the r-f is determined as a function of magnetic field strength. A typical curve of quenching *versus* magnetic field is shown in Fig. [7-6]. We have plotted in Fig. [7-7] the resonance magnetic fields for various frequencies in the vicinity of 10,000 Mc/sec." (The detailed paper states, "The resonance magnetic fields were located simply by taking the apparent peak in the case of sharp peaks, and by averaging the fields at half-amplitude for broad peaks.") "The theoretically calculated curves for the Zeeman effect are drawn as solid curves, while for comparison with the observed points, the calculated curves have been shifted downward by 1000 Mc/sec. (broken curves)." The improved agreement is obvious.

At this stage, "in view of the uncertainty introduced by the inhomogeneity of the field and distortion of the peaks due to falling off of the signal,[12] no attempt was made to obtain a 'best fit,' " and the value of the shift was quoted merely as 1000 ± 100 MHz. Further refinements permitted increased precision, to the point where these measurements became important data in determining the values of the fundamental constants of nature. The important fact, however, established by the first report, was the unambiguous existence of the effect.

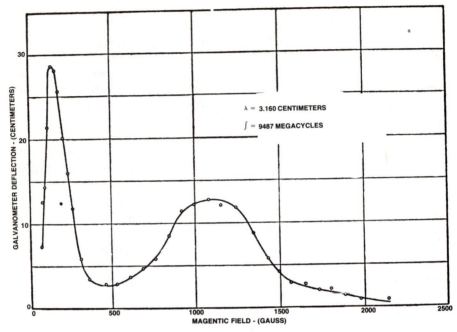

Fig. 7-6. A typical resonance curve. The left peak represents the transitions αc; the right, αb. [*Phys. Rev.* **72** (1947), p. 242, Fig. 1.]

As a postscript, it should be noted that 25 years after the work of Lamb and Retherford, it has become possible to resolve the fine structure of the H_α line by optical means, using a technique called *saturation spectroscopy*. This technique, as applied by T. W. Hänsch, I. S. Shakin, and A. L. Schawlow of Stanford University, involves the use of a tunable laser of very narrow linewidth. The laser beam is split into two parts, a strong saturating beam, which is periodically interrupted, and a weaker probe beam, both of which are passed in nearly opposite directions through the hydrogen sample. This eliminates the effect of the Doppler shift: Only atoms with zero velocity along the radiation path can interact with both beams, since nonzero velocity would shift the frequency upward for one beam and downward for the other. The atoms are excited to the $n = 2$ state by an electrical discharge. When the laser frequency is tuned to resonance with one of the frequencies of the H_α complex, either beam can then further excite the atoms to the $n = 3$ state, resulting in absorption from the beam. The presence of a resonance is signaled by a reduced absorption from the probe beam when the saturating beam is present, since the absorption by those atoms that can interact with both beams is saturated by the stronger beam.

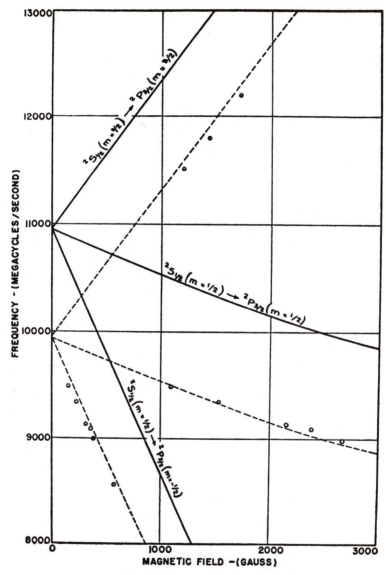

Fig. 7-7. Experimental values of resonance magnetic fields for various frequencies (circles), compared with values expected from Dirac theory (solid lines) and with the same values modified by the effect of an upward shift of the $2^2S_{1/2}$ level by 1000 MHz (dashed lines). "The plot covers only a small range of the frequency and magnetic field scale covered by our data, but a complete plot would not show up clearly on a small scale, and the shift indicated by the remainder of the data is quite compatible with a shift of 1000 Mc/sec." [*Phys. Rev.* **72** (1947), p. 242, Fig. 2.]

Notes

1. The value of L is conventionally indicated by a capital letter, with S denoting $L = 0$; P, $L = 1$; D, $L = 2$; and further correspondences that will not be needed here. The same letters in lower case designate the angular momentum of a single electron, which for hydrogen is the same as that of the atom; and thus are occasionally used that way in this work.

2. This is a standard spectroscopic notation in which the first number is the value of n; the superscript is $2S + 1$, where S is the total spin; the letter indicates the value of L, according to the scheme given in note 1; and the subscript gives the total angular momentum J.

3. It was Grotrian who introduced the sort of graphical representation of energy levels used in Fig. 7-1.

4. When the interaction of an atom with electromagnetic radiation is treated by means of quantum mechanics, it is found that radiation can be emitted or absorbed with substantial probability, not for arbitrary changes of atomic state, but only for certain ones, the restrictions being expressed in terms of changes in the quantum numbers of the atomic state. These transitions are called *allowed*, and the corresponding changes in quantum number are called *selection rules*.

5. See Chap. 6.

6. The mixing under the influence of an external electric field is called the *Stark effect*.

7. This is an effect brought about by the interaction of the magnetic moment with the external field, the energy being $\mu_z B$, where the z axis is along B. The component μ_z can be written as $g_J M_J$, where M_J is the projection of J along B; the coefficient g_J depends on J and L (and S, which is always $1/2$ for one-electron states).

8. Low-energy electrons traveling across a magnetic field are confined to helical paths around the field lines.

9. The FP54 is an amplifier tube specially designed to have a very high input impedance so as to respond measurably to very small currents.

10. More generally, each atomic level is split into a number of components equal to the smaller of $2I + 1$ and $2J + 1$, where I is the "spin" of the nucleus. For hydrogen, $I = 1/2$. This effect gives rise to hyperfine structure in optical spectra; cf. Chap. 6, p. 87.

11. A rectangular wave guide cannot transmit waves of frequencies below that for which the wavelength is twice the larger dimension. The wave guide used here is 0.400×0.900 in., so the cutoff frequency is 6550 MHz.

12. Even with no applied rf or dc quenching field, the detector signal decreases as the magnetic field increases, all other conditions remaining unchanged, because of the quenching due to the motional electric field (see p. 108). This field, and with it the amount of quenching, increases with the magnetic field.

Bibliography

The original report is Willis E. Lamb, Jr. and Robert C. Retherford, *The Physical Review* **72**, 241 (1947). The detailed series is Willis E. Lamb, Jr. et al., *The Physical Review* **79**, 549 (1950), 81, 222 (1951), **85**, 259 (1952), **86**, 1014 (1952), and **89**, 98 and 106 (1953). Also of interest is a paper reporting measurements of the same quantity in ionized helium: Willis E. Lamb, Jr. and Miriam Skinner, *The Physical Review* **78**, 539 (1950).

The first two papers of the detailed series are reprinted in full in *Molecular Beams: Selected Reprints. Vol. II, Atomic & Molecular Beam Spectroscopy* (American Institute of Physics, New York, 1965).

See also Lamb's Nobel prize lecture, in *Nobel Lectures*, vol. III, p. 286.

Chapter 8

The Magnetic Moment
of the Electron

In the early 1890s an outstanding physicist, Lord Kelvin, made a statement to the effect that the basic principles of physics had been completely established and all that was left to do was to "measure the next place of decimals." He was, of course, woefully wrong in his primary assumption: The succeeding decade alone witnessed the discovery of x rays and radioactivity, the demonstration that atoms are not indivisible, and the introduction of the quantum hypothesis. What is less widely appreciated is that important clues to a new understanding of nature can, in fact, be revealed by an improvement in the accuracy of measurement. This chapter describes a case in point: A highly precise determination of the magnetic moment of the electron yielded a value different by about a tenth of a percent from that predicted by theory. This discrepancy, like the level shift in hydrogen discussed in Chapter 7, played a part in stimulating the development of an improved theory of the interaction between the electron and the electromagnetic field. Because of the importance of this discovery, the man responsible for it, Polykarp Kusch of Columbia University, was awarded a Nobel Prize in physics in 1955.

The Bohr theory, of course, included the magnetic effects resulting from the electron's motion in its orbit, and incidentally provided a natural unit of magnetic moment: the *Bohr magneton*, $\mu_0 = e\hbar/2mc$, where e and m are the charge and mass of the electron, and c is the speed of light. The magnetic moment of an electron in an orbit of angular momentum $l\hbar$ was just $l\mu_0$. Consequently, when George Uhlenbeck and Samuel Goudsmit in 1925 introduced the concept of the electron spin and associated magnetic moment, it was natural for them to express the magnetic moment in that unit and to relate it also to the magnitude of the spin. However, they found that if the spin angular momentum was written as $s\hbar$, with $s = 1/2$, the intrinsic magnetic moment was not just $s\mu_0$, but $2s\mu_0$.

When one or more electrons are bound in an atom, the total resultant angular momentum is the vector sum of the individual contributions, both orbital and spin, and is expressed as $J\hbar$. The total magnetic moment is also a sum of the individual contributions. The measurable part of it is the component parallel to the vector **J**, and, by extension of the preceding arguments, it is desirable to express it as a multiple of J. If it were not for the extra factor 2 in the intrinsic magnetic moment, the total magnetic moment would be just $J\mu_0$. In actuality, there is an additional factor, which depends on how the individual contributions couple together and which is conventionally designated by g_J; thus $\mu_{\text{eff}} = g_J J \mu_0$. By analogy, one writes $\mu_{\text{intrinsic}} = g_s s \mu_0$; according to the hypothesis of Uhlenbeck and Goudsmit, then, one has $g_s = 2$. Soon after Uhlenbeck and Goudsmit made their proposal, Ernst Back and Alfred Landé measured μ_s and found that the value of g_s was indeed 2 to within 0.1%. When this value was found to be given as an inherent consequence of the relativistic theory of the electron developed by Dirac, it appeared to be firmly established.

However, in 1947, J. E. Nafe, E. B. Nelson, and I. I. Rabi applied Rabi's resonance molecular-beam method (Chap. 6) to a measurement of the hyperfine structure of the spectra of hydrogen and deuterium, that is, the splitting of an atomic level caused by the interaction between the total electronic magnetic moment and the magnetic moment of the nucleus. The values they obtained for the splitting of the lowest level in each of the two atoms differed from the theoretical values by about the same amount, roughly one quarter of one percent. Since the magnetic moment of the electron appears twice in the theoretical expression (once because it is the electronic moment that is involved in the interaction and once because the magnetic moment of the nucleus was most accurately known as a multiple of the electron moment), one explanation of the discrepancy, put forward quite hesitantly by Gregory Breit, was that the value of g_s for the electron is not exactly 2.[1] Kusch and Henry Foley therefore undertook to determine the value more directly and accurately. The results, announced first in a Letter to the Editor of *The Physical Review* in 1948 and in greater detail in an Article in the same journal later in the year, was that the value of g_s was not exactly 2, but $2 \times (1 + 1.19 \times 10^{-4})$.

The experimental method used was by this time standard; it was the way in which it was used and the way in which the results were interpreted that were ingenious.

"A deviation of the magnetic moment of the electron from the accepted value of one Bohr magneton could be detected by a precise measurement of the magnetic moment of an atom in a state in which the coupling of the electron spin with the orbital angular momentum is sufficiently well known. An absolute measurement of the magnetic moment requires a measurement of the Zeeman splitting of the zero field energy level in a known magnetic field. At the present time it is difficult to produce magnetic fields which are accurately known in terms of absolute standards and of sufficient magnitude to be useful in atomic beam determinations of the Zeeman splittings of energy levels. However, the frequencies of lines in the Zeeman spectrum of atoms (that is, the differences between atomic energy levels) may be determined by the use of readily available techniques to within one part in ten or twenty thousand, and where precision is limited by statistical errors, . . . a considerable improvement in precision may be obtained by a suitable repetition of observations. From measurements of the frequencies of Zeeman lines in two atomic states arising in either the same or different atoms but in the same constant magnetic fields, it is possible to deduce the ratio of the values of the gyromagnetic ratios[2] of the two states. If the spin and orbit vectors are coupled in the same way in the two states, the measured ratio yields no information about the fundamental g values of the electron. If the spin orbit coupling in the two states is different, however, the electron spin g value may be determined in terms of the orbital g value, provided only that suitable information is available, either on experimental or theoretical grounds, as to the validity of the assumed coupling. The principal limitations on such an experimental determination of the electron spin g value are the accuracy in the determination of the line frequency (limitations imposed by a frequency meter and by the line widths) and the stability and homogeneity of the magnetic field.

"In actual practice the single atomic level described above is split into two or more h.f.s. [hyperfine structure] levels because of the presence of the nuclear angular momentum. In such a case the interpretation of data on line frequencies becomes considerably more complicated."

The detailed theory of the dependence of atomic energy levels, and thus of transition frequencies, on the external magnetic field involves the application of quantum mechanics and is beyond the scope of this book. A semiclassical sketch can be given, however.

When an atom is in a magnetic field H, taken to be in the z direction, the interaction between the atomic electrons and the field is described by the expression

$$\mathcal{H} = g_L \mu_0 L_z H + g_S \mu_0 S_z H. \tag{8-1}$$

Here g_L and g_S are the orbital and spin gyromagnetic ratios, respectively; μ_0 is the Bohr magneton, $e\hbar/2mc$; and L_z and S_z are the z components of the orbital and spin angular momenta, respectively. If L_z and S_z were constants of the motion, the contribution to the energy would be obtained from Eq. (8-1) simply by substituting their actual values in each of the states involved; but this is practically never the case, and what must be substituted is the average value of each quantity. The average, in turn, can be expressed in terms of variables that are constants of the motion, or nearly so. For the kinds of atoms used—low to medium atomic number and only a small number of optically active electrons—the constants of the motion are L and S, in addition to J and J_z. The increment of energy is given by

$$W = (\alpha_L g_L + \alpha_S g_S)\mu_0 J_z H,$$

where

$$\alpha_S = [2J(J+1)]^{-1}[J(J+1) + S(S+1) - L(L+1)]$$

and

$$\alpha_L = [2J(J+1)]^{-1}[J(J+1) - S(S+1) + L(L+1)].$$

On the other hand, it is also given by

$$W = g_J \mu_0 J_z H,$$

so that

$$g_J = \alpha_L g_L + \alpha_S g_S.$$

"The ratio of the g values of two atomic states in the same or different atoms is:

$$g_{J_1}/g_{J_2} = [(g_L \alpha_{L_1} + g_S \alpha_{S_1})/(g_L \alpha_{L_2} + g_S \alpha_{S_2})],$$

in which it is assumed that the values of g_L and g_S are independent of the atomic state. If the fundamental gyromagnetic ratios differ from the conventional values by small amounts, then

$$g_S = 2(1 + \delta_S),$$

$$g_L = 1 + \delta_L, \tag{8-2}$$

and

$$g_{J_1}/g_{J_2} = [(2\alpha_{S_1} + \alpha_{L_1})/(2\alpha_{S_2} + \alpha_{L_2})]$$

$$+2[(\alpha_{S_1}\alpha_{L_2} - \alpha_{L_1}\alpha_{S_2})/(2\alpha_{S_2} + \alpha_{L_2})^2]\{\delta_S - \delta_L\}.$$

Thus if the constants α_{S_1}, α_{L_1}, α_{S_2}, α_{L_2} are known from the state of coupling of the atomic levels, the quantity $\{\delta_S - \delta_L\}$ can be determined from the ratio of the atomic g_J values. Clearly, no experiment of this type can distinguish between an effect produced by a small change from the previously accepted values of the spin or the orbital gyromagnetic ratios. . . .

"If hyperfine interactions were absent or entirely negligible the ratio of the frequencies of the Zeeman lines of two atomic states would give directly the ratio of the atomic g_J values. . . . In the atomic states which were studied in the present experiments and at the values of the magnetic field at which the lines were observed, the splitting of the energy levels into a hyperfine structure must be taken into account. The existence of this hyperfine structure complicates the analysis of the data from which the ratio of the g_J values is obtained; at the same time the possibility of observing a number of lines of different frequency resulting from hyperfine transitions within each atomic level gives a means of checking the self-consistency of data and improving the accuracy of results."

The interaction taking into account the effects of the nuclear moments is given by

$$\mathcal{H} = a\mathbf{I} \cdot \mathbf{J} + 2b\mathbf{I} \cdot \mathbf{J}(2\mathbf{I} \cdot \mathbf{J} + 1) + \mu_0 H(g_J J_z + g_I I_z).$$

Here $\mathbf{I}$ and $\mathbf{J}$ are quantum mechanical operators corresponding to the nuclear spin and the total electronic angular momentum, respectively, and g_I is the gyromagnetic ratio of the nucleus; a and b are constants characteristic of the particular atom; and $\mathbf{I} \cdot \mathbf{J}$ is a shorthand for $I_x J_x + I_y J_y + I_z J_z$. The first term represents the interaction between the electrons and the nuclear magnetic moment; the second, the interaction between the electrons and the nuclear electric quadrupole moment. The resulting contribution to the energy depends in a complicated manner on the state of the atom and is not readily expressible in terms of the constants of the motion as it was for the simpler case considered in Eq. (8-1). Rather, a set of equations is obtained for W/a, the ratio of the energy to the hyperfine coupling constant a, in terms of parameters $x = (g_J - g_I)(\mu_0 H/a)$ [or $(g_J - g_I)(\mu_0 H/2a)$ if the quadrupole terms need not be considered], $y = (g_J + g_I)(\mu_0 H/a)$, and $r = b/a$. The constants a and b, as well as g_I, had already been determined from other work.

"The quantities x and y always contain the factor $\mu_0 H/a$ (or $\mu_0 H/\Delta W)^3$ where a and ΔW are expressed in ergs. If we express all frequency measurements in terms of megacycles per second, then the quantity above becomes:

$$(\mu_0 H/ha) \times 10^{-6} = (eH/4\pi ma) \times 10^{-6} = H'/a,$$

or

$$H' = 1.3998H.$$

In all subsequent discussion we use the quantity H' as a measure of the field. . . .

"In the present series of experiments the direction of the oscillating magnetic field was perpendicular to the constant magnetic field. Thus the allowed transitions[4] are given, in weak field notation, by $\Delta F = \pm 1, 0$, and $\Delta m_F = \pm 1$.[5] Under very strong magnetic field conditions the nuclear and atomic angular momenta are decoupled, and the transitions may be classified according to $\Delta m_J = \pm 1$, $\Delta m_I = 0$ or $\Delta m_J = 0$, $\Delta m_I = \pm 1$. Whether or not a particular field is 'weak' or 'strong' for a particular atomic state depends on the value of x (or y) which corresponds to that field. This, in turn, depends on the value of ΔW or a.

"The decisions as to what magnetic field strength should be employed and which lines should be observed were made on the basis of the following considerations. Since the present experiments are directed toward the measurement of an atomic magnetic moment the lines to be observed should be selected to possess the greatest possible field sensitivity. Under very strong field conditions the field sensitivity of the transitions $\Delta m_I = \pm 1$ is so small that these lines are not useful for the purpose of this experiment. The transitions $\Delta m_J = \pm 1$ show adequate field sensitivity, but in most cases these lines occur at frequencies which are difficult to obtain experimentally. Under very weak field conditions the transitions

$\Delta F = 0$ are the most suitable for this experiment, as the frequencies of these lines are very nearly proportional to the field strength and are sufficiently field sensitive. Very weak fields ($H < 100$ gauss) could not be employed in this experiment because of the residual inhomogeneity of the field.

"The value of the magnetic field used in most of these experiments (~ 400 gauss) represents very weak field conditions for $^2P_{1/2}$ (In^{115}),[6] $^2P_{1/2}$ (Ga^{69}); for $^2S_{1/2}$ (Na^{23}) a considerable departure from weak field conditions appears, and for gallium in the $^2P_{3/2}$ state the field is very strong. Only a few lines in the spectrum of $^2P_{3/2}$ (Ga) of frequency less than 10^9 sec.$^{-1}$ permit a satisfactory determination of g_J. . . .

"In principle it is possible, from the observed frequencies of lines in the spectra of atoms in two different states, to calculate directly the ratio g_{J_1}/g_{J_2}. However, such a procedure is extremely laborious. We have, instead, calculated the quantity H' for each observed line. If a discrepancy in the value of H' occurs for atoms in two different states and under conditions for which H' is known to be identical, the discrepancy may be removed by an adjustment of the g_J values. This is evident since in the expressions for the energies of the levels, H' always occurs in the product $g_J H'$.[7] Suppose the assumed values of g_J are $g_{J_1}{}^0$ and $g_{J_2}{}^0$ for two different states. The corresponding values of H' are H_1' and H_2'. $\Delta H' = H_1' - H_2'$. Then:

$$g_{J_1}/g_{J_2} = (g_{J_1}{}^0/g_{J_2}{}^0)[1 + (\Delta H'/H')].$$

"The general procedures and instrumental requirements for the observation of lines in the radio frequency spectra of atoms have been discussed in a number of papers.[8] The molecular beam apparatus used in the experiments described [here] was originally designed as an apparatus for the study of the radio frequency spectra of molecules in very high magnetic fields. Accordingly, the deflecting fields are long and may be operated at high flux densities. . . . The magnetic field in which transitions occur is 48 cm long. . . . Unfortunately, the field is not entirely homogeneous, and the use of a large fraction of the length of the transition field for the observation of extremely field sensitive atomic lines broadens the lines excessively. In all of the present experiments, the effective length of the transition field was reduced to 2 cm by the expedient of reducing the length of the r-f field through which the beam passes. The deflecting fields were operated at a very low level of flux density. . . .

"For the purpose of consecutive measurement of lines in the radio frequency spectra of two different atomic species, such as gallium and sodium, a special oven chamber was constructed in which an oven containing gallium and another containing sodium were mounted on a platform attached to a ground joint which could be rotated from the outside of the apparatus. This arrangement permitted rapid interchange and adjustment of the two ovens.

"All frequency measurements were made on a . . . heterodyne frequency meter. . . . The meter may be used to determine frequencies to about one part in twenty thousand. . . . The uncertainties in frequency measurement imposed by the meter and by the line widths are statistical in character and the precision is improved by judicious repetition of observations." The wave meter was calibrated by comparison with signals from the National Bureau of Standards radio station.

"All experimental data considered in the experiments discussed [here] are measurements of the frequencies of spectral lines. No knowledge of the magnetic field in which transitions occur is required, though it is necessary that transitions resulting in lines whose frequencies are to be compared, occur in the same magnetic field. In practice, a magnetic field is not entirely constant, and the rate of drift of the field depends on a number of factors, of which the principal two are the condition of the storage cells which supply the exciting current to the magnet and the temperature stability of the d.c. circuit and the magnet. A drift is observed by noting a change in the frequency of a line with time and a very good correction for drift of field may be made by alternately measuring the frequencies of each of two lines to be compared and then, by suitable graphical or other methods, determining the frequencies of each of the lines at any arbitrary instant of time.

"To reduce the time interval between observations on successive lines, and hence minimize the effects of a drifting field, all observations were made on the basis of a single frequency reading for each line"; that is, the full resonance curve was not traced out, but rather the observer recorded for each line only the value corresponding to maximum reduction in beam intensity, obtained by means suitable to each particular line.

After some preliminary work to establish that frequencies could indeed be determined to a precision of one part in 20,000, several intercomparisons were made. The detailed description of one of these, that between sodium and gallium, gives the general picture of the procedure.

"Three lines in the sodium spectrum which have the greatest sensitivity to field were measured alternately with two lines in the spectrum of gallium." The lines were given identifying designations as indicated in Table 8-I. "The data are given in Table [8-II]. The time observation is indicated in the first column, and in the other

Table 8-I. Identification of lines used in the comparison between sodium and gallium. All sodium lines are within the $^2S_{1/2}$ state; all gallium lines are within the $^2P_{1/2}$ state. [Adapted from *Phys. Rev.* **74** (1948), p. 257, Table V.]

Designation	$(F, m_F) \longleftrightarrow (F', m_F')$
Na*I*	$(2, -2) \longleftrightarrow (2, -1)$
Na*II*	$(1, 0) \longleftrightarrow (1, -1)$
Na*III*	$(2, 0) \longleftrightarrow (2, -1)$
Ga*I*	$(1, 0) \longleftrightarrow (1, -1),$ Ga^{69}
Ga*II*	$(2, 0) \longleftrightarrow (2, -1),$ Ga^{69}

columns are given the observed line frequencies in megacycles per second. Directly below the observed frequency is recorded, in parentheses, the value of H' calculated on the basis of the assumption that $g_J(^2S_{1/2}) = 2$ and $g_J(^2P_{1/2}) = 2/3$. The quantity which appears directly after the value of H' is the difference between the observed H' and that calculated from a least squares solution of the data.

"The data are presented graphically in Fig. [8-1]. An independent least squares solution of the variation of H' with time for sodium and for gallium gives very nearly the same rate of drift of field in both cases, as expected.[9] A combined least squares solution for both sets of lines, where the rate of drift of field is the same in both cases, yields the result

$$\text{for Na:} \ H' = 531.319 - 0.002475t,$$
$$\text{for Ga:} \ H' = 530.036 - 0.002475t, \qquad [(8\text{-}3)]$$

where the time, t, is measured in minutes from 6:00. The difference between the observed and calculated values is indicated in the table. . . . The difference in the value of H' for Na and that for Ga is $531.319 - 530.036 = 1.283$. To this we assign the arbitrary precision value of ± 0.030, which is the sum of the mean deviation of the values of H' yielded by Eq. [(8-3)] from those calculated for the individually observed Na lines and of the corresponding quantity for Ga. . . .

"On the basis of the assumption that $g_J(^2S_{1/2}) = 2$ and that $g_J(^2P_{1/2}) = 2/3$, the ratio of the apparent field as calculated from Na to that as calculated from Ga is 1.00242 ± 0.00006. . . . To satisfy the condition that the magnetic field is the same for a measurement of the frequencies of the gallium and sodium lines, we must set:

$$g_J(^2S_{1/2} \text{ Na})/g_J(^2P_{1/2} \text{ Ga}) = 3(1.00242 \pm 0.00006),$$

Table 8-II. The observed frequencies and values of H' of several lines in the spectrum of sodium and gallium. [*Phys. Rev.* **74** (1948), p. 258, Table VII.]

Time	Na*I*	Na*II*	Na*III*	Ga*I*	Ga*II*
6:30				91.093 (529.97) + 0.00	
6:40					90.319 (529.93) − 0.01
6:50				91.087 (529.93) + 0.01	
7:05			260.32 (531.16) + 0.00		
7:05		261.17 (531.14) − 0.02			
7:15	417.33 (531.15) + 0.02				
7:40				91.060 (529.78) − 0.01	
7:40					90.295 (529.79) + 0.00
7:58			260.27 (531.02) − 0.01		
8:02		261.12 (531.01) − 0.01			
8:14	417.17 (531.00) + 0.01				
8:28				91.038 (529.65) − 0.02	
8:35					90.275 (529.68) + 0.02
8:45			260.21 (530.89) + 0.00		
8:49		261.05 (530.85) − 0.05			
9:10	417.04 (530.88) + 0.03				
9:10	417.05 (530.89) + 0.04				
9:26				91.014 (529.52) − 0.01	
9:26					90.248 (529.52) − 0.01

where it is assumed that the deviation of the g_J values from their nominal values is small."

Altogether, three independent intercomparisons were made on four different atomic states. Although, as noted earlier (page 125), deviations from the expected values could be due to discrepancies in either g_L or g_S, Kusch and Foley chose to interpret them entirely in

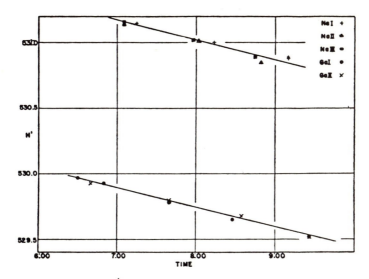

Fig. 8-1. Values of H' as functions of time, calculated from the observations tabulated in Table 8-I, showing the effect of the drift of the magnet current. [*Phys. Rev.* **74** (1948), p. 259, Fig. 2.]

terms of modifications in g_S—that is, in terms of a nonzero value for the quantity δ_S introduced in Eq. (8-2). The results are shown in Table 8-III. They acknowledged that "the difference between the value of δ_S which results from the $^2P_{3/2}(\text{Ga}) - {}^2P_{1/2}(\text{Ga})$ experiment and that which results from the $^2P_{1/2}(\text{Ga}) - {}^2S_{1/2}(\text{Na})$ experiment is probably real"—even though it is just at the limit of the assigned uncertainties—but they felt that even this might be accounted for by "small deviations of the properties of the atomic systems from the simple description implicit in the theory by means of which the experimental data has been reduced." In any case, they emphasized that "the numerical results are independent of the precise values of any of the fundamental atomic constants."

Kusch and Foley regarded the agreement of the values of δ_S as "very strong evidence in support of the hypothesis that the funda-

Table 8-III. Observed ratios of atomic g_J values and the corresponding values of δ_S. [*Phys. Rev.* **74** (1948), p. 262, Table X.]

Experimental ratio	δ_S
$g_J(^2P_{3/2}\ \text{Ga})/g_J(^2P_{1/2}\ \text{Ga})$ = 2(1.00172 ± 0.00006)	0.00114 ± 0.00004
$g_J(^2S_{1/2}\ \text{Na})/g_J(^2P_{1/2}\ \text{Ga})$ = 2(1.00242 ± 0.00006)	0.00121 ± 0.00003
$g_J(^2S_{1/2}\ \text{Na})/g_J(^2P_{1/2}\ \text{In})$ = 2(1.00243 ± 0.00010)	0.00121 ± 0.00005

mental spin gyromagnetic ratio does, in fact, differ from the accepted integral value." They considered "several perturbations of the electronic states involved . . . which in principle could bring about deviations of the atomic g values from the values given by the . . . coupling formula" and concluded that these could at most give corrections "well within the experimental uncertainty." They conclude, "We write $g_S = 2(1.00119 \pm 0.00005)$ as the best value of the spin gyromagnetic ratio obtainable from the present experiments."

The case was not entirely closed, even though the same issue of *The Physical Review* that contained Kusch and Foley's second Letter also carried a Letter by Julian Schwinger who reported theoretical calculations that gave a value $g_S = 2 \times 1.0016$. Nevertheless, as Kusch expressed it in his Nobel lecture, "The importance of the observation of the anomalous magnetic moment of the electron is in part in the demonstration that the procedures of quantum electrodynamics are, in fact, satisfactory in formulating a description of nature."

As a postscript to this chapter, it is interesting to note that the anomalous value of g_s for the electron was almost discovered over a decade earlier. In 1934, L. E. Kinsler at the California Institute of Technology was studying the Zeeman effect as a means for evaluating the charge-to-mass ratio of the electron, e/m. The deduction of e/m from the measured wavelength differences involves, in addition to a high-precision measurement of the magnetic field, a knowledge of the way in which the individual electron spins and orbital angular momenta are coupled. However, there are certain quantities or combinations of quantities that are independent of the nature of the coupling. One such combination is provided for by a relation called the *g-sum rule*, which states that the sum of the values of g_J for all the states with a given value of J formed from a given combination of single-electron states is independent of the coupling. Kinsler undertook to test this rule, using two states in neon that involved five electrons in $2p$ states and one in a $3s$ state. Neon conforms fairly well to the conditions mentioned earlier in the chapter under which L and S are approximately constants of the motion, and thus the two states can be labeled according to that scheme: They are $^1P_1$ and $^3P_1$. If the equations given in that discussion are used to calculate the values of g_J, the results are

$$g_J(^1P_1) = g_L, \quad g_J(^3P_1) = \frac{1}{2}(g_L + g_S), \qquad (8\text{-}4)$$

so that the g-sum rule gives

$$\sum g_J = \frac{3}{2} g_L + \frac{1}{2} g_S.$$

Kinsler's measurements gave

$$g_J(^1P_1) = 1.0350 \pm 0.0007,$$

$$g_J(^3P_1) = 1.4667 \pm 0.009.$$

The discrepancies from the values given in Eq. (8-4) are substantial but are presumably to be ascribed to the imperfectness of the L-S description. The g-sum rule is another matter:

$$\sum g_J = 2.5017.$$

Kinsler simply added the errors on the individual values and quoted a resultant error of 0.0016, so that the discrepancy was regarded as insignificant. However, Kusch has remarked that this is "perhaps . . . a needlessly large uncertainty." If the individual errors were independent, a better overall uncertainty would have been the square root of the sum of the squares, or 0.0011, making the discrepancy large enough to be taken seriously. It corresponds to a value of 0.0017 for δ_S, as compared with Kusch's determination of 0.0012.

Even if Kinsler had believed his result, however, the tone of his paper suggests that he would have ascribed it to the failure of the g-sum rule rather than to an anomalous value of g_S. On such minor things does the course of events depend!

Notes

1. After all, the proton itself seemed—and still seems—to be well described by the Dirac theory except for *its* magnetic moment, which corresponds to a value for g_S of 2.7.

2. The gyromagnetic ratio is the ratio of the magnetic moment in units of the Bohr magneton (or, for a nucleus, the nucleon magneton) to the angular momentum in units of h, in other words, any of the factors g introduced earlier.

3. For the cases where the quadrupole term need not be considered, the hyperfine splitting in zero magnetic field is denoted by ΔW and is equal to $2a$.

4. See note 4 of Chap. 7.

5. The symbol F is the conventional notation for the total angular momentum of the atom, $\mathbf{F} = \mathbf{I} + \mathbf{J}$; and m_F denotes its component along the z axis.

6. See note 2 of Chap. 7 for an explanation of the notation for atomic states.

7. "It is true that a term $g_I H'$ also occurs. However, in [all cases but one] the value of g_I has been determined in each case in terms of the ground state of the atom in which the nucleus occurs. The term $g_I H'$ may then be written as

$(g_I/g_J)g_JH'$ where the ratio g_I/g_J is a quantity independent of any assumption as to g_J." In the one exceptional case, "the term is so small that uncertainties in g_I have no appreciable effect on our results."

8. See, for example, Chap. 6.

9. The tabulation, in addition to showing qualitatively both the drift and the presence of a real $\Delta H'$, also displays the need for time to interchange and readjust the ovens and to change the frequency ranges.

Bibliography

The original reports of this work are P. Kusch and H. M. Foley, *The Physical Review* **72**, 1256 (1947); H. M. Foley and P. Kusch, ibid. **73**, 412 (1948); and P. Kusch and H. M. Foley, ibid. **74**, 250 (1948). See also Kusch's Nobel lecture, in *Nobel Lectures*, vol. III, p. 298; and, particularly for a sketch of subsequent developments, P. Kusch, *Physics Today* **19**, No. 2, 23 (1966).

Chapter 9

The Transistor

Probably no single development of modern physical science has touched so many people's lives so directly as has the transistor. With its advantages over the electron tube—the most obvious being that it is much smaller, it operates at lower voltages so that it does not require bulky power supplies, and it does not have a filament which takes time to warm up and from which the heat must be dissipated—the transistor revolutionized electronic communication devices, as well as making practical the extensive development of high-speed, high-capacity computers. In regard to the latter, additional important features of the transistor are the low amount of energy required per bit of information processed[1] and its extremely long operational life. Its invention was truly a landmark, and it is small wonder that a Nobel prize for physics was awarded in 1956 to the men primarily responsible: John Bardeen, Walter Brattain, and William Shockley.

However, in contrast to most of the other events described in this book, the story of the invention of the transistor is not the story of a single, fairly elaborate experiment. Instead, as will be told in this chapter, it was merely one step in a rather extensive program of research on semiconductors, carried out by a closely knit team that included not only physicists but a circuit expert and a physical chemist and that had close ties with a metallurgical group. As Bardeen described it in his Nobel lecture, "The general aim of the program was to obtain as complete an understanding as possible of semiconductor phenomena, not in empirical terms, but on the basis of atomic theory." Indeed, the Nobel prize was not just for the invention of the transistor, but for the whole research program. Nevertheless, Bardeen noted, "Aside from intrinsic scientific interest, an important reason for choosing semiconductors as a promising field in which to work, was the many and increasing applications of them in electronic devices, which, in 1945, included diodes, varistors and thermistors. There had long been a hope of making a triode, or an amplifying device with a semiconductor." Thus, while the work was not aimed primarily at the development of a device, the possibility was always in mind.

To be able to follow the story, it is necessary to know something of the nature of semiconductors and the reason for their properties. Significant research on semiconductors dates back at least as far as 1833, when Michael Faraday discovered that the conductivity of a semiconductor (he was working with silver sulfide) increases as the temperature rises, instead of decreasing as does the conductivity of a metal. By the end of the nineteenth century, three other important properties were discovered: the production of an electromotive force when a semiconductor used as one electrode in an electrolytic cell is illuminated[2]; the increase in conductivity of a semiconductor by illumination; and the rectifying properties of a contact between a metal and a semiconductor.[3] It should be noted that two methods for greatly altering the conductivity of a semiconductor, by temperature and by illumination, were discovered quite early. The transistor effect was to add a third, control of the conductivity by the presence of a current.

In the 1920s, the photovoltaic effect and the rectification began to be commerically developed; and it began to appear that these were surface effects, while photoconductivity and the negative temperature coefficient of resistance were effects related to the bulk of the material. By the end of that decade, it was recognized that conductivity depended both on the number of charge carriers per unit volume and on their mobility, which is the ratio between the drift velocity imparted to a carrier by an electric field and the magnitude of that field. In the early 1930s, measurements of the Hall effect[4] in semiconductors revealed that both of these factors had quite a different character in semiconductors from what they had in metals. One the one hand, the carrier density in metals is almost constant, not only from one metal to another, but also as the temperature changes. In semiconductors, on the other hand, it varies from one specimen to another, even of the same material, and increases strongly with temperature; and it is much smaller—by orders of magnitude—than in metals. The carrier mobility in both metals and semiconductors varies significantly from one material to another and decreases as the temperature rises; in metals this is the controlling temperature effect, but in semiconductors it is overwhelmed by the increase in carrier density.

The Hall effect also permits determination of the sign of charge of the carriers or, at least, of those that dominate the conduction process. In some semiconductors, at temperatures that were not too high, these were found to be positive; and for some semiconductors that were compounds rather than elements (for example, copper oxide or zinc oxide), the sign of the dominant carrier often appeared to be correlated with a minute departure of the composition from that given by the chemical formula. Thus if there was a deficiency of a few parts per million of Cu in Cu_2O, the carriers were positive; if there was a deficiency of O in ZnO, they were negative. At high temperatures (room temperatures and above), however, the conductivity was always due predominantly to negative carriers, whose density might still vary from one substance to another but not between specimens of a single substance.

The growth of understanding of the behavior of electrons in solids based on quantum mechanics permitted the development, by A. H. Wilson in 1931, of a model of a semiconductor which correlates many of these properties. The model, schematized in Fig. 9-1, is based on the fact that when atoms coalesce to form solids, the discrete energy states of the atomic electrons broaden into essentially continuous bands separated by "gaps" of energy values inaccessible to the electrons. The electrons in a fully occupied band cannot contribute to an electric current in a solid, but those in an incompletely filled band can. Metallic conduction occurs when the highest occupied band is only partially filled, so that a large number of electrons can contribute to the current; this case is represented in part (a) of Fig. 9-1. In an ideal, pure semiconductor, the highest occupied band is completely filled at zero temperature, and the next highest band is completely empty.[5] However, the gap E_g between these bands is fairly small, of the order of an electron volt. The probability that an electron will be thermally excited across the gap is roughly proportional to $\exp(-E_g/kT)$,[6] where k is Boltzmann's constant, 8.617×10^{-5} eV/K, and T is the absolute temperature. Thus, for temperatures that are not too low, there will be a significant number of electrons[7] excited to the conduction band, where they can contribute to the conductivity; see part (b) of Fig. 9-1. It is this that accounts for most of the high-temperature behavior described earlier. In addition, according to quantum theory, the absence of an electron from a normally occupied state (called a *hole*) is equivalent to the presence of a positive charge. The holes left

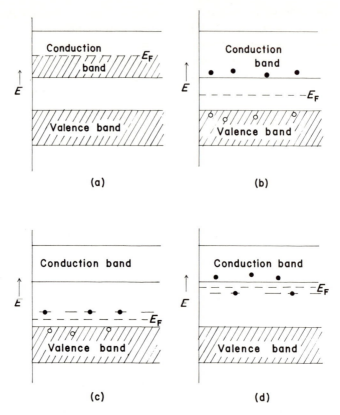

Fig. 9-1. (a) Schematic representation of the energy states in a metal. The valence band is shown shaded [in parts (b)-(d) as well], as is the mostly occupied portion of the conduction band. (b) Energy states in an intrinsic semiconductor. Small circles in the valence band represent holes left by thermal excitation of electrons to the conduction band, where they are shown as dots. (c) Energy states in a defect (acceptor) semiconductor. The electrons excited out of the valence band now go to acceptor states, represented by the short lines. (d) Energy states in an excess (donor) semiconductor. The electrons in the conduction band are excited out of the donor states, shown as short lines. In all four parts, the Fermi level is shown as E_F.

in the valence band by thermal excitation also contribute to the high-temperature, or *intrinsic*, conductivity, but their mobility is lower than that of the electrons.

The low-temperature behavior is explained by the model in terms of imperfections in the lattice. In copper oxide, for example, with an excess of oxygen, each unoccupied copper position in the lattice would mean a deficiency of one electron from the number needed to fill the valence band—a hole. At very low temperatures, the hole is bound to the vacant site; but the energy necessary to free it is much smaller than the gap energy, of the order of 0.1 eV, so that it can contribute to the conductivity of low temperatures. An alternative way to view this process is to say that this sort of defect produces an energy state for electrons just a little above the top of the valence band; this state is unoccupied at very low temperatures, but as the temperature is increased an electron from the valence band may be thermally excited to fill it, leaving the hole in the valence band free to contribute to the conductivity. This is illustrated in part (c) of Fig. 9-1. The extra state, which is highly localized, is referred to as an *acceptor* state. Acceptor states may also be produced by the presence of impurity atoms that do not contribute the normal complement of bonding electrons, for example, a trivalent atom, such as gallium, in the crystal of a tetravalent element, such as silicon or germanium. In a similar manner, an imperfection that results in an excess of electrons—an excess of metal in a metal oxide or sulfide, or a pentavalent impurity such as arsenic or phosphorus in a tetravalent element—gives rise to an energy state just a little below the bottom of the valence band, called a *donor* state, as shown in part (d) of Fig. 9-1; and relatively little thermal energy is needed to excite the electron out of one of these states and into the conduction band.

Also indicated in each part of Fig. 9-1 is a special value of the energy, the Fermi energy E_F. This value is defined, roughly, as the energy such that the probability of an electron occupying a state at that energy is 1/2 (or would be 1/2 if such a state existed). In a metal, the Fermi energy is equal to that of the highest-energy state occupied at zero temperature. In an intrinsic semiconductor, it is approximately at the middle of the gap; in an impurity semiconductor, it is displaced from that position by an amount and in a direction that depends on the relative values of the concentration n_e of free electrons and n_h of holes: upward if $n_e > n_h$, downward if $n_e < n_h$.[8] The importance of the Fermi energy is that it constitutes the electrochemical free energy in thermodynamics and must, therefore, have the same value on both sides of a phase boundary or an interface for thermodynamic equilibirum. When two different solids are in contact, this coincidence of values is brought about by one of the solids assuming a higher potential (the contact potential difference) than the other.

Thermal excitation is not the only possible means of raising electrons from one level to another. Photons can also provide the necessary energy if their frequency is high enough that their energy $h\nu$ is at least equal to the excitation required. Thus excitation involving impurity states can be produced with moderately long wavelengths; excitation across the full gap, only with somewhat shorter wavelengths. It should be noted, however, that at room temperature essentially all acceptor levels are already full and all donor levels empty, so that almost all photoexcitation involves the production of electron-hole pairs.

Thus, the model accounts reasonably well for those features of semiconductor behavior that are related to bulk properties but says virtually nothing about the surface phenomena. One approach to accounting for these attempted to relate them to the fact that a double charge layer exists at the surface of a metal. This fact had been known already early in the century; the layer is necessary in order to account for the difference in potential between the interior and exterior of a metal as revealed, for example, in the thermionic work function. J. Frenkel of the University of Petrograd (now Leningrad) gave the following qualitative explanation for such a layer in *The Philosophical Magazine* in 1917:

"Let us imagine a surface passing through the outermost nuclei and call it, for the sake of brevity, the surface of the body. One half of the electrons rotating around these nuclei will remain outside this surface along with a lot of other electrons belonging to nuclei which are situated within the surface at a distance not exceeding the radius of the largest electronic orbits. We may identify this radius, corresponding to the 'valency electrons,' with the atomic radius r. Thus a layer of thickness r within the surface will be positively charged owing to part of the electrons which belong to its nuclei remaining outside the surface and forming there a negative layer of the same thickness r." After the development of quantum theory, calculations by Frenkel and, later, by Bardeen put this explanation on a firmer and more quantitative basis.

According to this concept, when two dissimilar metals are brought into virtual contact, the double-layer charges redistribute themselves, assisted by some electron flow from one metal to the other, until the Fermi levels coincide; see part (a) of Fig. 9-2. The net remaining double layer accounts for the contact potential difference. It should be noted that because the conduction electrons in metals are essentially free, this redistribution and flow causes no significant change in the state of affairs in the interior of the metal. The case of a contact between a metal and a semiconductor or between two semiconductors is expected to be quite different. Consider for definiteness the contact between a metal and a donor semiconductor, as shown in part (b) of Fig. 9-2. Before equilibrium is established, the electrons in the donor states are at energies corresponding to or above vacant states in the metal, leaving a net positive charge inside the semiconductor in a region near the surface. In this case, however, there is no significant replenishment from the interior, and the energy levels there are unaltered. The overall effect is, as shown, the

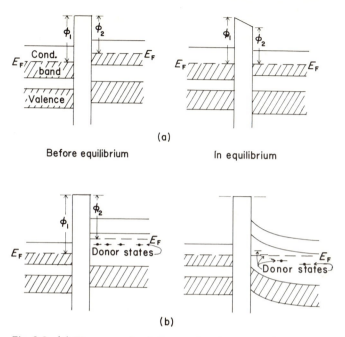

Fig. 9-2. (a) The energy-level diagram for the contact between two dissimilar metals. (b) The same for the contact between a metal and a donor semiconductor. In both parts, the ϕ's are the work functions.

production of a space-charge layer which forms a barrier against the passage of electrons between the conduction bands of the two substances.

There is also, of course, a barrier remaining in the case of contact between metals. But it is very narrow, of the order of 10^{-8} cm, and because of the tunnel effect, it offers no appreciable resistance to current flow. In contrast, for metal-semiconductor contact the barrier layer is of the order of 10^{-4} cm thick, and no significant tunneling current is possible.[9] At ordinary temperatures, however, there can be sufficient thermal excitation to permit conduction over the barrier. The barrier is asymmetrical, and its height is different according to the direction of the applied voltage. N. F. Mott in England, B. Davydov in the U.S.S.R., and especially Walter Schottky in Germany utilized these ideas in 1939 to formulate a theory of the rectification process.

About 1940, Russell S. Ohl of Bell Telephone Laboratories, bothered by the lack of consistency in the properties of silicon as used in the old "cat's-whisker" radio detectors, approached the chemists and metallurgists with the suggestion that they might make the material more uniform by making it purer. Such proposals were

given greater urgency by the importance of point-contact semiconductor diodes in radar and other electronic devices in World War II.
Those primarily responsible for the work were J. H. Scaff, H. C.
Theuerer, and E. E. Schumacher. They were able not only to
produce materials of considerably higher purity than had been
previously available, but also to control the type of impurity; they
soon recognized that the conductivity was mainly due to small traces
of impurities. It was during this period that Ohl introduced the
terminology "n-type" or "p-type" for silicon in which the easy flow
of current at a point contact occurred when the silicon was negative
or positive, respectively. Scaff, Theuerer, and Schumacher also
discovered that impurities tend to remain in the molten material
as an ingot is solidified,[10] but the tendency is not equally strong for
the two types—those producing n-type and those producing p-type
material. Thus if an ingot were allowed to freeze progressively from
one end to the other, one end would be n-type and the other p-type.
Scaff, Theuerer, and Schumacher could then isolate the impurities.
They showed that elements from the fifth column of the periodic
table acted as donors and produced n-type material; those elements
from the third column were acceptors and produced p-type.

The group at Bell Laboratories, with William Shockley as its
head, was formed in 1946. They decided to concentrate their
attention on silicon and germanium rather than on the more widely
used copper and zinc oxides, partly because of the greater simplicity
of elements as opposed to compounds, but in large measure because
of the development of materials technology just described.[11]

By about this time it had become clear that the existing theory
of rectification was inadequate. It predicted a dependence of
rectifying ability on the work function of the metal that was not
found experimentally; it predicted a contact potential difference
between n-type and p-type silicon that was not found; and it predicted that a contact between two semiconductors of the same
material, one p-type and one n-type, should be a good rectifier,
whereas actually it behaved like two opposing rectifiers.

There was also another piece of evidence which at first seems
unrelated. It occurred to Shockley that it should be possible to
modify the space-charge layer by means of an electric field; if the
sample were very thin, so that the space-charge region were a
significant fraction of the thickness, a current flowing parallel to the
surface should be modified in turn, resulting in amplification. The
experiment was tried, but a report of it was never published because
the effect was far smaller than anticipated.

As Brattain later tells it, "[T]he group as a whole slowly
realized that the results were all of a piece, and it was Bardeen who

successfully explained them all by applying to this problem the concept of surface states[12]; that is, that the electrons could be trapped at the semiconductor surface, and that the semiconductor was in equilibrium with its surface before any electrical contact was made with it. This, of course, implied a space charge layer in the surface of the semiconductor equal and opposite to the charge trapped on the surface. Consequently, the electrostatic potential change between the interior of the semiconductor and the surface which was necessary for rectification was a property of the semiconductor and its surface—independent of the metal contact."

This proposal not only explained the lack of dependence of rectifier behavior on the nature of the metal; it also explained the failure to observe the field effect, as the electrons in surface states would effectively shield the interior from the external field.

The essential qualitative features of the theory can be understood with reference to Fig. 9-3.

"The lowest state of the conduction band and the highest state of the filled band of the semi-conductor are indicated, with an energy gap ϵ_g. It is assumed that the distribution of surface states is such that the surface states give no net charge if the states are filled to an energy ϵ_0 below the conduction band. Since the Fermi level cuts the surface above the level determined by ϵ_0, the surface will be negatively charged, this charge resulting from electrons in states between ϵ_0 and the Fermi level. The picture applies to an excess semi-conductor. [A footnote at this point says, "The case of an excess semi-conductor seems easier to visualize than that of a defect semi-conductor. All results derived for one case, of course, apply to

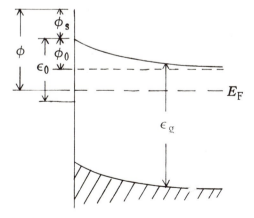

Fig. 9-3. The energy-level diagram for states at and near the surface of a semiconductor, including surface states. [Adapted from *Phys. Rev.* **71** (1947), p. 722, Fig. 3.]

the other with obvious changes in signs of the charges."] In the body of the semi-conductor, the Fermi level is an energy ζ below the conduction band. The space charge region extends for an approximate distance l into the semi-conductor, giving a potential energy rise ϕ_0 at the surface. . . .

"The amount and extent of the space charge inside the free surface is determined by the density of surface levels. For zero external field, the positive space charge raises the potential at the surface by an amount just sufficient to give a compensating negative surface charge. The larger the positive space charge region, the larger is ϕ_0, and the smaller is the negative surface states charge. For some ϕ_0 the two will be equal in magnitude. This is the equilibrium value."

Even though the correctness of this theory was strongly indicated by its agreement with already known facts, other experiments to verify it were desirable. One such experiment was based on the prediction that the number of surface states per unit energy should be roughly proportional to the concentration of donor or acceptor states; consequently, the equilibrium value of ϕ_0, and with it the contact potential of the semiconductor, should also depend on the donor or acceptor concentration. Verification of this prediction was reported by Brattain and Shockley in a Letter to the Editor in *The Physical Review*.

"The contact potentials of several N and P-type silicon surfaces have been measured. The respective number of donor and acceptor impurities in each sample were determined by measuring the electrical conductivity and Hall constant.[13] A correlation was found between impurity concentration and contact potential. Each surface was ground flat and then sand-blasted lightly with 180 mesh silicon carbide. The contact potential was then measured in air, in vacuum after heat treatment to 400°C, in high purity N_2 and finally in air again. The results are given in Table [9-]I. The reference surface was platinum, which probably had a work function between 4 and 5 volts. The data given are averages of several runs and should be accurate to approximately 0.02 volts.

". . . On the basis of the surface state picture one would expect the contact potential between N and P-type silicon to increase as the respective impurity concentration was increased approaching 1.2 volts [the width of the energy gap in silicon] as a limit. The data in Table [9-]I shows that after heat treatment in vacuum the contact potential difference between the two types does increase and the difference between P-type silicon (5.7×10^{20} acceptors) and N-type silicon (1.9×10^{20} donators) is the order of 0.6 volts. From these data one can estimate the density of surface states in silicon to be approximately 10^{14} per volt per cm^2."

Table 9-I. Contact potential, in volts, for a series of silicon semiconductor samples after various forms of surface treatment. [*Phys. Rev.* **72** (1947), p. 345, Table I.]

Type	Impurity N/cm^3	In air after sandblast	In vacuum after heat treatment	After letting in N_2	After exposure to air
P	5.7×10^{20}	+0.31	−0.27	−0.19	+0.07
P	1.5×10^{20}	+0.35	−0.18	−0.10	+0.13
P	6.5×10^{18}	+0.30	−0.10	−0.09	+0.17
P	3.1×10^{17}	+0.34	+0.04	+0.14	+0.28
N	6.9×10^{18}	+0.32	+0.16	+0.27	+0.34
N	2.3×10^{19}	+0.37	+0.27	+0.35	+0.39
N	1.9×10^{20}	+0.37	+0.30	+0.37	+0.37

The theory also indicated the importance of studying the surface properties directly. One such property (related to the photovoltaic effect) is the change of contact potential on illumination. Immediately following the Letter by Brattain and Shockley was one by Brattain reporting a study of that effect.

"On the basis of the surface state picture one expects a double layer at the free surface of a semiconductor, the excess charge in the surface states being compensated by an equal and opposite space charge in the semiconductor. At low temperatures it should take an appreciable time for this equilibrium to be established between the surface states and the space charge region. Consequently, electrons or holes excited by absorption of light near the surface should upset this equilibrium making the surface state charge more positive for N-type semiconductors and more negative for P-type semiconductors, thus changing the contact potential when the surface is exposed to light. The experiment suggested has been done on silicon surfaces of both N and P type and on an N-type germanium surface. The contact potential changes due to light exposure at approximately 120°K were +0.12 volts for N-type silicon, −0.08 volts for P-type silicon, and +0.02 volts for N-type germanium. The changes have the predicted sign. The change appeared to be instantaneous on exposure of the surface to light. The return to the equilibrium condition in the dark appeared to have a time constant of the order of a few seconds. No change in contact potential with exposure to light was found at room temperature."

Bardeen in his Nobel lecture describes the method used in these experiments: "Apparatus used by Brattain to measure contact potential and change in contact potential with illumination is shown in Fig. [9-4]. The reference electrode, generally platinum, is in the form of a screen so that light can pass through it. By vibrating the

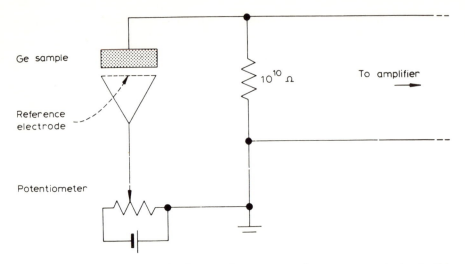

Fig. 9-4. Schematic diagram of apparatus used to measure contact potentials and their variation under illumination. [*Nobel Lectures in Physics*, vol III (Elsevier Scientific Publishing Company, Amsterdam, 1964), p. 335, Fig. 9. Used by permission.]

electrode, the contact potential itself can be measured by the Kelvin method.[14] If light chopped at an appropriate frequency falls on the surface and the electrode is held fixed, the change with illumination can be measured from the alternating voltage developed across the condenser."

Brattain next, as he described it later, tried "to measure the change of potential at the germanium or silicon surface as a function of temperature. Condensation of moisture from the air on the cold semiconductor surface interfered with this experiment. As a result it was decided to try immersing the system in an insulating liquid. The apparatus had been arranged to measure contact potential and photo emf's, and when the liquids were tried large changes in photo emf's were observed. Some of the liquids tried (such as water) were not strictly insulating but were electrolytes. When I was showing these phenomena to [another member of the group, physical chemist R. B.] Gibney he suggested varying the potential between the semiconductor surface and the reference electrode. When using an electrolyte we could make the photo emf very large by this means. By changing the sign of the potential we could make the photo emf go through zero and change sign. It was recognized that this was, in essence, Shockley's field effect. By using the electrolyte we could vary the space charge layer and potential inside the semiconductors at the surface. [Bardeen, in his Nobel lecture, said, "Evidently ions piling up at the surface created a very large field which penetrated through the surface states." Brattain has since stated that hindsight indicates, rather, a change of the surface states by the applied potential.]

"The results were presented to the group as a whole and one morning one or two days later Bardeen came into my office with a suggested geometrical arrangement to use this effect to make an amplifier. I said let's go out in the laboratory and do it. We covered a metal point with a thin layer of wax, pushed it down on a piece of *p* type silicon that had been treated to give it an *n* type surface. We then surrounded the point with a drop of water and made contact to it [see Fig. 9-5]. The point was insulated from the water by the wax layer. We found as expected that potentials applied between the water and the silicon would change the current flowing from the silicon to the point.[15] Power amplification was obtained that day!"

The action of the device in this form is as follows: The metal point is biased in the "reverse" direction, that is, the direction of small current flow. Part of the current that does flow consists of electrons passing through the *n*-type surface layer. The application of a negative potential to the electrolyte reduces the number of electrons in the layer by changing the surface states (the field effect) and thereby reduces the current to the metal point. Only a very small current flows in the electrolyte, and, thus, current and power amplification were achieved, but not voltage amplification. Bardeen has noted that essential to the success of the experiment was the use of material whose bulk electrical properties were very good, and the restriction of electron flow to the surface layer by modification of

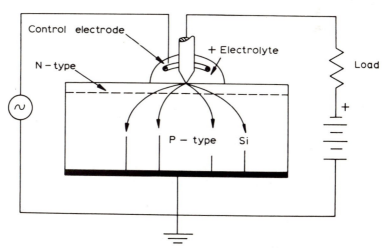

Fig. 9-5. Schematic diagram of the first semiconductor amplifier arrangement, as used by Bardeen and Brattain. [*Noble Lectures in Physics*, vol III (Elsevier Scientific Publishing Company, Amsterdam, 1964), p. 336, Fig. 10. Used by permission.]

the material itself, making an n-type layer at the surface of p-type material, rather than by means of an evaporated film in which the mobility would have been much lower.

Brattain's account goes on: "Bardeen suggested trying this on *n* type germanium, and it worked even better. However, the water drop would evaporate almost as soon as things were working well, so at Gibney's suggestion we changed to glycol borate, which hardly evaporates at all. Another problem was that amplification could be obtained only at or below about 8 Hz. We reasoned that this was due to the slow action of the electrolyte. Optimum results were obtained with a dc negative bias on the electrolyte when using *n* type germanium. Under these conditions we noticed an anodic oxide film being formed under the electrolyte. We decided to evaporate a spot of gold on such a film and, using the film to insulate the gold from the germanium, use the gold as a field electrode to eliminate the electrolyte. The film was formed, the glycol borate washed off, and the gold spot with a hole in the middle for the point was evaporated. When this was tried, an electrical discharge between the point and the gold spoiled the spot in the middle; but by placing the point around the edge of the gold spot, a new effect was observed. In washing off the glycol borate, we had inadvertently washed off the oxide film which was soluble in water. The gold had been evaporated on a freshly anodized germanium surface. When a small positive potential was applied to the gold, holes flowed into the germanium surface greatly increasing the flow current from the germanium to the point negatively biased at a large potential!"

After all this preliminary work, none of which was published, the formation of a useful device was almost anticlimactic. The appropriate experimental arrangement had to be worked out; Brattain, in an interview[16] taped in 1964 for the Center for the History of Physics at the American Institute of Physics, has described an important part of that arrangement.

"After discussions with John Bardeen we decided that the thing to do was to get two point contacts on the surface sufficiently close together, and after some little calculation on his part, this had to be closer than 2 mils.[17] The smallest wires that we were using for point contacts were 5 mils in diameter. How you get two points 5 mils in diameter sharpened symmetrically closer together than 2 mils without touching the points, was a mental block.

"I accomplished it by getting my technical aide to cut me a polystyrene triangle which had a small narrow, flat edge and I cemented a piece of gold foil on it. After I got the gold on the triangle, very firmly, and dried, and we made contact to both ends of the gold, I took a razor and very carefully cut the gold in two at the apex of the triangle. I could tell when I had separated the gold. That's all I did. I cut carefully with the razor until the circuit opened and put it on a spring and put it down on the same piece of germanium that had been anodized but standing around the room now for pretty near a week probably. I found that if I wiggled it just right so that I had contact with both ends of the gold that I could make one contact an emitter and the other a collector, and that I had an amplifier with the order of magnitude of 100 amplification, clear up to the audio range."

Success was achieved on 23 December 1947 and reported by Bardeen and Brattain in a Letter to the Editor in *The Physical Review* the following July; a detailed discussion of the physical principles and electrical characteristics was given in an Article in *The Physical Review* in April 1949. The following description is from the Letter.

"A three-element electronic device which utilizes a newly discovered principle involving a semiconductor as the basic element is described. It may be employed as an amplifier, oscillator, and for other purposes for which vacuum tubes are ordinarily used. The device consists of three electrodes placed on a block of germanium as shown schematically in Fig. [9-6]. [See also the two photographs in Fig. 9-7.] Two, called the emitter and collector, are of the point-contact rectifier type and are placed in close proximity (separation ~ .005 to .025 cm) on the upper surface. The third is a large area low resistance contact on the base.

"The germanium . . . is an *N*-type or excess semiconductor with a resistivity of the order of 10 ohm cm. . . . Both tungsten and phosphor bronze points have been used. . . ."

"Each point, when connected separately with the base electrode, has characteristics similar to those of the high back-voltage rectifier.[18] Of critical importance for the operation of the device is the nature of the current in the forward direction. We

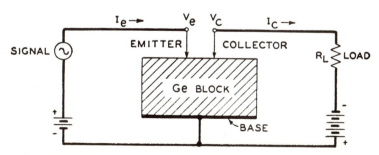

Fig. 9-6. "Schematic of semi-conductor triode." [*Phys. Rev.* **74** (1948), p. 230, Fig. 1.]

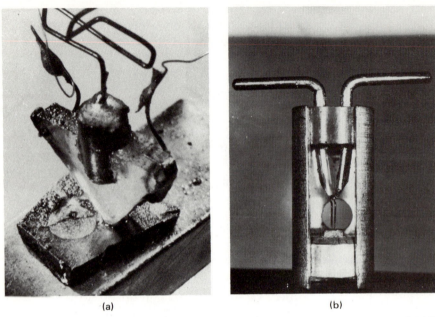

(a)　　　　　　　　　　　　　　　　(b)

Fig. 9-7. (a) A photograph of the original transistor. [*Phys. Teacher* 6 (1968), p. 112, Fig.5.] (b) "Microphotograph of a cutaway model of a transistor." [*Phys. Rev.* **75** (1949), p. 1210, Fig. 2.]

believe . . . that there is a thin layer next to the surface of *P*-type (defect) conductivity. As a result, the current in the forward direction with respect to the block is composed in large part of holes, i.e., of carriers of sign opposite to those normally in excess in the body of the block.

"When the two point contacts are placed close together on the surface and d.c. bias potentials are applied, there is a mutual influence which makes it possible to use this device to amplify a.c. signals. A circuit by which this may be accomplished is shown in Fig. [9-6]. There is a small forward (positive) bias on the emitter, which causes a current of a few milliamperes to flow into the surface. A reverse (negative) bias is applied to the collector, large enough to make the collector current of the same order or greater than the emitter current. The sign of the collector bias is such as to attract the holes which flow from the emitter so that a large part of the emitter current flows to and enters the collector. While the collector has a high impedance for flow of electrons into the semiconductor, there is little impediment to the flow of holes into the point. If now the emitter current is varied by a signal voltage, there will be a corresponding variation in collector current. It has been found that the flow of holes from the emitter to the collector may alter the normal current flow from the base to the collector in such a way that the change in collector current is larger than the change in emitter current. Furthermore, the collector, being operated in the reverse direction as a rectifier, has a high impedance (10^4 to 10^5 ohms) and may be matched to a high impedance load. A large ratio of output to input voltage, of the same order as the ratio of the reverse to the forward impedance of the point, is obtained. There is a corresponding power amplification of the input signal. . . .

"Using the circuit of Fig. [9-6], power gains of over 20 db [that is, a factor over 100] have been obtained. Units have been operated as amplifiers at frequencies up to 10 megacycles."

The fact that the change in collector current is actually larger than the change in emitter current means that it is not simply a matter of holes injected by the emitter flowing into the collector. The detailed Article gives the following explanation: "The fact that the collector current may actually change more than the emitter current is believed to result from an alteration of the space charge in the barrier layer at the collector by the hole current flowing into the junction. The increase in density of space charge and in field strength make it easier for electrons to flow out from the collector, so that there is an increase in electron current."

This explanation involves two concepts that had gradually begun to evolve in the course of the experiments. One is that a change in the density of minority carriers, even though small in an absolute sense, may have a profound effect on the overall properties of the material. The other is that the essential feature of an active circuit element (such as an amplifier) is the presence of two phase boundaries—between metal and semiconductor, between two types of semiconductor, or even between metal and vacuum—close enough together that nonequilibrium phenomena due to current flow across one boundary can influence the current flow across the other, and having electrical connections to the three phases separated by the boundaries. The second of these took longer to become clear, however, and its elucidation was to lead to important further developments.

It is to be recalled that Bardeen ascribed the success of the work with electrolytes to the presence of a modified layer on the surface of the semiconductor, with the electron flow restricted to that layer. However, this interpretation was early called into question, especially when it was learned that, as stated in the extended Article, "Good transistors can be made with surfaces prepared in the usual way for high back-voltage rectifiers provided that the collector point is electrically formed.[19] Such surfaces exhibit no measurable surface conductivity." The implication that the modified surface layer was unimportant was tested in several ways, the most clear-cut of which was proposed by Shockley and independently carried out by John N. Shive: The emitter and collector points were placed in contact with opposite sides of a thin piece of germanium, and the device gave transistor action comparable with that of the original arrangement.

In attempting to devise experiments bearing on this point, Shockley was led to consider structures of three layers, one of, say, *n*-type material between two of *p*-type. He had thought of such structures before, though not primarily in connection with amplifiers; but this time the pieces fell into place. The diffusion of minority carriers *into* the central portion could be controlled by a potential applied to that portion, but once in, they would almost certainly diffuse *through*. This is the essence of the operation of *p-n-p* and *n-p-n* junction transistors. The detailed theory was published in 1949, and the first successful device was constructed in late April 1950.[20] Since then, junction transistors have largely supplanted the original point-contact type.

Notes

1. The amount of energy required is comparable, it is reported, with that used by the neurons in the brain.

2. This effect was discovered by Edmond Becquerel, father of the Henri Becquerel who discovered radioactivity.

3. More properly, it is the fact that the ratio between current and voltage is not a constant, as in Ohm's law, but depends on the sign as well as the magnitude of the voltage.

4. It is defined as follows: When a current flows in a conductor perpendicular to a magnetic field, a potential difference appears between the sides of the conductor, that is, in a direction perpendicular to both the current and the magnetic field.

5. The highest band that is completely filled at zero temperature in an ideal crystal is called the *valence band*; the band next above that is called the *conduction band*.

6. $\exp(x) = e^x$, where $e = 2.31828\ldots$

7. For room temperature ($T = 300$ K) and $E_g = 1$ volt, the value of the exponential is roughly 10^{-17}. However, since there are of the order of 10^{23} electrons/cm^3 of solid in a filled band, this means that about 10^6 electrons/cm^3 are excited into the conduction band.

8. The exact value is $E_F = E_{F0} + \frac{1}{2}kT \ln(n_e/n_h)$, where E_{F0} is the value it would have if the semiconductor were intrinsic; "ln" stands for natural logarithm.

9. The layer can be made even larger in alkali halides, and in 1938 R. Hilsch and R. W. Pohl actually succeeded in incorporating an electrode into a

potassium bromide crystal and using it as a control grid in a triode. The device, however, was usable only at frequencies of the order of a hertz.

10. This tendency was utilized more fully some 15 years later, in the technique known as *zone refining*, in which a zone of melting is moved along a rod of material. The impurities tend to be carried along in the molten zone and are thus segregated at one end. By means of repeated passes, the technique produces some of the purest materials yet available.

11. By no means was all of this work done at Bell. Important contributions, scientific as well as technological, were made by groups at Purdue University and Massachusetts Institute of Technology and in England.

12. This concept had been introduced several years earlier by I. Tamm (who later shared a Nobel prize for his studies concerning the emission of electromagnetic radiation by fast charged particles passing through matter), and had been subsequently studied by many people, including Shockley in particular.

13. This is the factor of proportionality between the transverse electric field produced by the Hall effect (see note 3) and the product of the current density and the magnetic field: $E = R_H jH$. It is inversely proportional to the carrier concentration.

14. This is the method used by Millikan for measuring contact potentials in connection with his experiments on the photoelectric effect; see *Crucial Experiments*, p. 82.

15. The "current flow" mentioned in this sentence is electron flow; the conventional current flow is in the opposite direction.

16. Cited in Charles Weiner, *IEEE Spectrum* 10, 24 (1973).

17. 1 mil = 0.001 in.

18. A high back-voltage rectifier is one that can withstand a high voltage in the reverse, or low-conducting, direction.

19. This is the process of passing pulses of high current through the point. The actual nature of what happens is not understood.

20. Junction transistors could not actually be constructed until after G. K. Teal and J. B. Little had developed the technology of growing large single crystals of germanium.

Bibliography

The earliest paper that seems to have a direct bearing on the development of the transistor is John Bardeen's paper on surface states: *The Physical Review* 71, 717 (1947). This was followed by the two Letters reporting verification of predictions of the theory: W. H. Brattain and W. Shockley, *The Physical Review* 72, 345 (1947); W. H. Brattain, ibid. The Letter announcing the transistor itself is J. Bardeen and W. H. Brattain, *The Physical Review* 74, 230 (1948); the detailed Article is J. Bardeen and W. H. Brattain, *The Physical Review* 75, 1208 (1949).

An extensive history of semiconductor research is given by G. L. Pearson and W. H. Brattain, *Proceedings of the Institute of Radio Engineers* **43**, 1794 (1955), and a more limited and more informal history of the development of the transistor by Walter H. Brattain, *The Physics Teacher* **6**, 109 (1968). Other aspects of the history are given by Charles Weiner, *IEEE Spectrum* **10**, 24 (1973), and by William Shockley, in *Proceedings of Conference on Public Need and the Role of the Inventor, June 11-14, 1973, Monterey, California*, National Bureau of Standards Special Publication No. 388 (U.S. Government Printing Office, Washington, D.C., 1974), chap. 9.

See also the Nobel prize lectures: John Bardeen, in *Nobel Lectures*, page 318; William Shockley, ibid., p. 344; Walter H. Brattain, ibid., p. 377.

Chapter 10

Disproof of a Conservation Law

It is fair to say that most of the effort in physics is directed at understanding how material systems change in the course of time, under the influence of either external agents or the mutual interactions of the parts of the system. This being so, it may seem a little incongruous that among the most important laws of physics are the conservation laws, statements attesting that under very broad conditions certain quantities do not change at all. The reason for this importance is that they are insensitive to the details of the process being considered; as one author puts it, "[I]n this very incompleteness lies the value of the law. A whole host of different processes are covered by the same quantitative statement."[1] Clearly, the discovery of a new conservation law would be a matter of great note; correspondingly, the abandonment of an old one is an action not taken without the strongest evidence. This chapter describes the gathering of such evidence, and the motivation for doing so, in the case of the law of conservation of a quantity called *parity*.

The concept of parity is intimately related to the distinction between left- and right-handedness and to mirror symmetry or the lack of it. The natural world is full of examples of left-right asymmetry, ranging from such gross (and apparently unimportant) features as the placement of body organs or the direction of coiling of certain mollusk shells to the much more subtle (but biologically significant) phenomenon of optical activity[2] of many organic molecules. But all such examples appear to be accidental and not to reflect any inherent asymmetry in the laws of nature. Thus, any optically active substance can exist in two forms, whose molecules are mirror images of each other and which rotate polarization in opposite directions.[3] While natural organisms can utilize, for example, only the levorotatory form[4] of sugar, there is no reason to doubt that organisms could exist that would utilize the dextrorotatory form (and the mirror-image form of every other optically active biological substance) and which would be otherwise identical to known organisms. In other words, the laws of classical physics (and chemistry) are invariant under a reflection of the coordinate system.

Invariance principles are ordinarily related to conservation laws. The law of conservation of energy, for example, is a manifestation of the invariance of physical laws under a shift of time. In a sense, this relationship holds for reflection invariance also. Consider a string stretched between fixed supports. The natural modes of vibration of the string can be classified according to their properties under reflection through the center of the string: The odd harmonics are unchanged, while the even harmonics change sign (in the second harmonic, for example, the left half is displaced upward when the right half is displaced downward; reflection interchanges these). If the string is given an initial displacement that is unaltered by reflection, the resultant motion contains only odd harmonics and remains unchanged under reflection. There is an important difference, however, between reflection invariance and other types of invariance. The continuous symmetries—rotation and space and time translation—are related to dynamical quantities which are themselves continuously variable, and a dynamical system can be characterized by values of these variables. It is these values that are constrained by the conservation law. In contrast, the only property that can be related to reflection is evenness or oddness of the coordinate description. But a classical system may be neither even nor odd; and even if it is one of the two, the continuous nature of classical science makes inconceivable the idea of a change to the other—*natura non facit saltus.*

With the advent of quantum theory, reflection symmetry took on a new aspect. Now the state of a dynamical system was to be described by a single function of the coordinates of the particles composing the system, the so-called wave function. Any such function, if it does not already have specially simple properties under reversal of the coordinates, can be expressed as the sum of two functions that do: one which is unchanged ("even") and one which merely reverses its sign ("odd"). This in itself is nothing new; the same is true of, say, the function describing the configuration at any instant of the string considered earlier. But in quantum theory, each of the two functions represents a state of the system. Moreover, quantum theory permits discontinuous changes associated with transitions, so that a system originally described by a function of one of the two special forms might conceivably change to one described by the other. In 1927, however, Eugene Wigner noted that the only interaction then known, electromagnetism, was of such a mathematical form that this could not happen, and that furthermore a pure state of a system would be described by a function that had one of the two special forms. Thus parity was elevated to the status of a quantum number.

It is important to note that if a system is composed of distinct parts, such as an atom and a radiation field, the overall wave function can be expressed as a product of parts, one for each subsystem. The parity of the total system is the product of the parities of the parts, and it is the overall parity that is to be conserved. An atom initially in a state of even parity can undergo a transition to a state of odd parity (or vice versa) by emitting a photon into a state of odd parity. Thus the law of conservation of parity proved quite valuable to the understanding of the spectroscopy of many-electron atoms and molecules. By the mid-1950s, it was accorded the same nearly sacrosanct status as were the classical conservation laws.[5]

In the years prior to 1956, this special status was an essential element in a puzzle regarding the behavior of two of the recently discovered elementary particles. The particles were known as the τ and the θ. As far as could be determined, they both had the same mass (about 985 times that of the electron) and the same mean life (about one hundred-millionth of a second); but the τ decayed into three pions, the θ into two pions. It seemed very strange that two particles should be so similar in mass and lifetime and yet exhibit different modes of decay.[6] It would have been most appealing to regard them as merely two different decay modes of a single particle, but here is where the law of parity conservation interfered: An analysis by R. H. Dalitz showed that the two-pion state has even parity and the three-pion state has odd parity, and a parent particle that presumably has a definite parity, whichever it might be, cannot decay into both.

In April of 1956, two theorists, T. D. Lee of Columbia University and C. N. Yang of the Princeton Institute for Advanced Study, proposed one way out of the dilemma. Their scheme was known as "parity doubling"; its essential feature was the assumption that every elementary particle of odd "strangeness"[7] occurs in two forms, of opposite parities but otherwise identical. They presented their scheme again at a biennial International Conference on High Energy Physics at the University of Rochester that summer.

In the audience was Richard Feynman, a theorist from the California Institute of Technology who was renowned for his part in the reformulation of quantum electrodynamics.[8] Feynman had as his roommate during the conference an experimentalist, Martin Block, and on the first night Block had suggested to Feynman that perhaps parity was simply not conserved in the "weak" interactions, the class responsible for the decay of the θ and the τ. The two of them had discussed the matter for several nights and concluded that Block's question was not an unreasonable one; and in the discussion following Yang's presentation, Feynman brought it up on the floor. Yang replied that he and Lee had considered the idea but had not reached any firm conclusions. In the course of the discussion, Wigner—who, it will be recalled, had developed the idea of parity conservation in the first place—also conceded that perhaps it did not hold for weak interactions.

It is hard to doubt that Lee and Yang were encouraged by such interest. In any case, they proceeded to investigate how well founded the law of conservation of parity actually was. As was mentioned earlier, the law was originally deduced for electromagnetic interactions, and there it was unassailable; Lee and Yang found that it had also been verified for the very strong interaction that produces binding of the nucleus. But as far as Lee and Yang could discover, all previous experiments on processes governed by the weak interaction had been of such a kind that they were insensitive to whether the interaction led to conservation of parity. The law was simply assumed because it had been so well established before any theoretical developments regarding the weak interaction had been even conceivable.

Lee and Yang therefore proposed, in an Article in *The Physical Review* late in 1956, that "the present theta-tau puzzle may be taken as an indication that parity conservation is violated in weak interactions." They cautioned, however, "This argument is . . . not to be taken seriously because of the paucity of our present knowledge concerning strange particles. It supplies rather an incentive for an examination of the question of parity conservation." Lee and Yang accordingly went on to analyze a number of possible experiments which would test their suggestion, describing the kind of results that would be obtained if parity were in fact not conserved.

While most physicists probably felt that the Lee-Yang proposal was too drastic to have much chance of being correct, few if any could argue that it should not be tested. Two groups immediately took up the challenge, and a third started somewhat later. Their results, reported in three Letters in *The Physical Review* (received by the editor within an interval of two days) early in 1957, completely confirmed the conjecture; as a result, Lee and Yang shared the 1967 Nobel prize for physics.

One of the groups was sparked by a colleague of Lee's at Columbia, C. S. Wu. This remarkable woman already had a world-wide reputation for her excellent work on beta decay, the type of radioactivity in which an electron is emitted by the nucleus; and one of the kinds of experiments that Lee and Yang had discussed had to do with beta decay, which is mediated by the weak interaction. As stated in the Letter reporting this experiment, "In beta decay, one could measure the angular distribution of the electrons coming from the beta decays of polarized nuclei.[9] If an asymmetry in the distribution between θ and $180° - \theta$ (where θ is the angle between the orientation of the parent nuclei and the momentum of the electrons) is observed, it provides unequivocal proof that parity is not conserved in the beta decay." In other words, suppose the nuclei are oriented with their spins pointing up; then if the number of electrons emitted into the upper hemisphere is different from the number emitted into the lower, it is evidence that parity is not conserved.

It was known that many kinds of nuclei could be polarized by an external magnetic field.[10] However, the effect is useful only at very low temperatures; otherwise the thermal motion of the nuclei destroys the alignment. At the time, there was only one institution in the United States with the capability of providing the low-temperature environment for the kind of arrangement that was needed: the National Bureau of Standards, where Ernest Ambler, who had been a member of the first group to achieve nuclear polarization by the desired method, was working in the section headed by R. P. Hudson. Accordingly, Wu arranged for a collaboration with a group there consisting of Ambler, Hudson, and two nuclear physicists, R. W. Hayward and D. D. Hoppes.

The nuclide chosen for the study was ^{60}Co, practically ideal for several reasons. It emits both beta and gamma rays; it was known to be polarizable, with the degree of polarization measurable by the anisotropy of the gamma radiation, which tends to be emitted more in the polar direction than in the equatorial plane; and it can be easily incorporated into a salt, cerium magnesium nitrate, which can be brought to near absolute zero by adiabatic demagnetization.[11]

However, as the group noted in their Letter, "To apply this technique to the present problem, two major difficulties had to be overcome. The beta-particle counter should be placed *inside* the demagnetization cryostat, and the radioactive nuclei must be located in a *thin surface* layer and polarized." These requirements arise from the inability of beta particles to penetrate any substantial thickness of matter. "The schematic diagram of the cryostat is shown in Fig. [10-]1." A photograph is shown in Fig. 10-2.

"To detect beta particles, a thin anthracene crystal . . . is located inside the vacuum chamber about 2 cm above the Co^{60} source. The scintillations are transmitted through a glass window and a

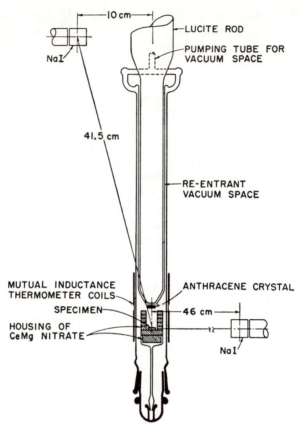

Fig. 10-1. Schematic diagram of the lower portion of the cryostat used in the polarized ^{60}Co experiment. [*Phys. Rev.* **105** (1957), p. 1413, Fig. 1.]

Lucite light pipe 4 feet long to a photomultiplier . . . which is located at the top of the cryostat. The Lucite head is machined to a logarithmic spiral shape for maximum light collection." The researchers checked to make sure that this did not seriously affect the resolution of the system. "The stability of the beta counter was carefully checked for any magnetic or temperature effects and none were found. To measure the amount of polarization of Co^{60}, two additional NaI gamma scintillation counters were installed, one in the equatorial plane[12] and one near the polar position. The observed gamma-ray anisotropy was used as a measure of polarization and, effectively, temperature. . . . Specimens were made by taking good single crystals of cerium magnesium nitrate and growing on the upper surface only an additional crystalline layer containing Co^{60}. . . . The thickness of the radioactive layer used was about 0.002 inch. . . . Upon

Fig. 10-2. Experimental setup of the polarized ^{60}Co experiment. In the fore-
ground is the glass cryostat swung out of the cooling magnet. The ^{60}Co source
is located in the lower part of the cryostat. The gamma-ray counters in the
equatorial plane are shown at the left. The huge magnet used in paramagnetic
cooling is shown in the background with poles widely open [Photo and cap-
tion courtesy of Dr. C. S. Wu.].

demagnetization, the magnet is opened and a vertical solenoid is raised around the lower part of the cryostat." It is this solenoid that provides the polarizing field. "The whole process takes about 20 sec. The beta and gamma counting is then started. The beta pulses are analyzed with a counting interval of one minute, and a recording interval of about 40 seconds." The procedure was to continue the counting and recording repeatedly as the sample temperature rose.

"A large beta asymmetry was observed. In Fig. [10-3] we have plotted the gamma anisotropy and beta anisotropy *vs* time[13] for polarizing field pointing up and pointing down. The time for disappearance of the beta asymmetry coincides well with that of gamma anisotropy. The warm-up time is generally about 6 minutes, and the warm counting rates are independent of the field direction. The observed beta asymmetry does not change sign with reversal of the demagnetization field, indicating that it is not caused by remanent magnetization in the sample."

The authors recognized the possibility that the effect might be produced by causes other than failure of parity conservation. "The double nitrate cooling salt has a highly anisotropic *g* value." That means that its magnetic properties vary with direction relative to the axes of the crystal lattice. "If the symmetry axis of a crystal is not set parallel to the polarizing field, a small magnetic field will be produced perpendicular to the latter. To check whether the beta asymmetry could be caused by such a magnetic field distortion, we allowed a drop of $CoCl_2$ solution to dry on a thin plastic disk and cemented the disk to the bottom of the same housing. In this way the cobalt nuclei should not be cooled sufficiently to produce an appreciable nuclear polarization, whereas the housing will behave as before. The large beta asymmetry was not observed." They also investigated "possible internal magnetic effects on the paths of the electrons as they find their way to the surface of the crystal," and convinced themselves that these were not significant. Thus, while they were still preparing to carry out "more rigorous experimental checks" (which, as it turns out, were never published), the evidence for nonconservation of parity was strong.[14]

At that time, it was a frequent practice of many of the physicists at Columbia to gather on Fridays for "Chinese lunch" under the guidance of T. D. Lee, who is something of a gourmet. At one of these lunches, Lee reported a piece of news he had just learned: that the Columbia–National Bureau of Standards experiment was giving positive results. Among those present was Leon Lederman, who, with a graduate student, Marcel Weinrich, was studying some aspects of muon behavior with the aid of the cyclotron at the university's new Nevis Laboratory. Realizing that he was in a position to make an in-

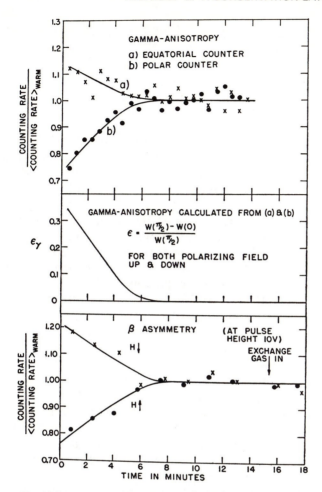

Fig. 10-3. Top: actual counting rates of the gamma counters, normalized to the average counting rate for a warm sample (for which the rates are equal). Middle: resulting anisotropy coefficient, the ratio of the difference between equatorial and polar rates to the equatorial rate; note that as the sample warms up, the anisotropy disappears, indicating loss of polarization. Bottom: beta counting rates for the two field directions, showing the marked asymmetry. [*Phys. Rev.* **105** (1957), p. 1414, Fig. 2.]

dependent test of the Lee-Yang proposal, Lederman quickly arranged for modification of the experimental setup for that purpose. In addition to Weinrich and himself, he included in the group Richard L. Garwin, whose exceptional creativity in instrumentation was to contribute much to the experiment.

Their approach, as pointed out in the Letter reporting their results, was based on the suggestion by Lee and Yang "... that confirmation [of their hypothesis] should be sought in the study of the successive reactions

$$\pi^+ \rightarrow \mu^+ + \nu, \qquad\qquad\qquad\qquad ([10\text{-}]1)$$

$$\mu^+ \rightarrow e^+ + 2\nu. \qquad\qquad\qquad\qquad ([10\text{-}]2)$$

They have pointed out that parity nonconservation implies a polarization of the spin of the muon from stopped pions in (1) along the direction of motion[15] and that furthermore, the angular distribution of electrons in (2) should serve as an analyzer for the muon polarization. . . .

"By stopping, in carbon, the μ^+ beam formed by forward decay in flight of π^+ mesons inside the cyclotron, we have performed the meson experiment, which establishes the following facts:

"I. A large asymmetry is found for the electrons in (2), establishing that our μ^+ beam is strongly polarized.

"II. The angular distribution of the electrons is given by $1 + a\cos\theta$, where θ is measured from the velocity vector of the incident μ's. We find $a = -\frac{1}{3}$ with an estimated error of 10%.

"III. In reactions (1) and (2), parity is not conserved. . . .

"The experimental arrangement is shown in Fig. [10-4]. The meson beam is extracted from the Nevis cyclotron in the conventional manner. . . . The positive beam contains about 10% of muons which originate principally in the vicinity of the cyclotron target by pion decay-in-flight. Eight inches of carbon are used in the entrance telescope to separate the muons, the mean range of the . . . pions being ~5 in. of carbon. This arrangement brings a maximum number of muons to rest in the carbon target."

The next stage is discussed in more detail in Weinrich's Ph.D. thesis. "In order to investigate the angular distribution of decay positrons, one could rotate a counter telescope . . . about the stopping target and so sample the distribution function at a number of angles. A simpler and more profitable method was employed: to sweep the distribution function across a fixed counter system. This was accomplished by the application of a vertical magnetic field at right angles to the muon polarization direction. The accompanying magnetic moment results in a gyroscopic precession [of the muon spins] in the horizontal plane, thus displacing the counters, respectively, by an angle:

$$\theta = \omega t = \frac{-\mu H}{s\hbar} t$$

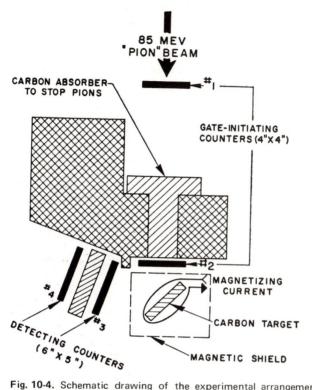

Fig. 10-4. Schematic drawing of the experimental arrangement used in the parity experiment at the Nevis Cyclotron Laboratory. [*Phys. Rev.* **105** (1957), p. 1415, Fig. 1.]

where:

μ = magnetic moment of the muon
s = spin of the muon[16]
H = magnitude of the magnetic field

and the sign indicates the direction of rotation in the conventional manner, that it is counterclockwise for both positive H (up) and positive moment. . . .

"The necessary magnetic field was obtained by winding a coil around the target, and passing a suitable current through it." This method was part of Garwin's contribution.[17]

"The detection apparatus consisted of two counter telescopes. One, the beam detector, contained two scintillation counters; the second, the electron telescope, consisted of either two or three counters, placed at an angle of $100°$ with respect to the beam. Fig. [10-4] shows the arrangement for [two] counters.

"The first scintillator in the beam was a $4'' \times 4'' \times 1/2''$ plastic scintillator viewed by a ... photomultiplier tube. Counter No. 2 placed $20''$ behind was only $1/2''$ thick, and of the same cross-section. The pulses from each counter were first amplified in a fast limiting amplifier, then put in coincidence with a resolving time of about 10^{-8} seconds. ... The coincidence pulses were then amplified and ... fed into a discriminator followed by a pulse shaping stage which ... [gave] a standard pulse, ... referred to as the 12 pulse. The positrons emanating from the target were detected by a telescope of plastic scintillators (labelled 3, 4, and 5). The scintillator is 6 $1/2''$ high, $5''$ wide, $1/2''$ thick, and is viewed through a tapered lucite light pipe by a ... photomultiplier. The coincidence arrangement is the same as for the beam detectors, with more amplification needed for the counter pulses."

The remaining procedure was adequately described in the published version:

"The stopping of a muon is signalled by a fast 1-2 coincidence count. The subsequent beta decay is detected by the electron telescope 3-4 which normally requires a particle range > 8 g/cm^2 ($\sim$25-Mev electrons) to register. ... Counting rates are normally $\sim$20 electrons/min in the μ^+ beam ... with background of the order of 1 count/min.

"In the present investigation, the 1-2 pulse initiates a gate of duration $T = 1.25$ μsec. This gate is delayed by $t_1 = 0.75$ μsec and placed in coincidence with the electron detector. Thus the system counts electrons of energy > 25 Mev which are born between 0.75 and 2.0 μsec after the muon has come to rest in carbon. Consider now the possibility that the muons are created in [the reaction $\pi^+ \rightarrow \mu^+ + \nu$] with large polarization in the direction of motion. If the gyromagnetic ratio is 2.0, these will maintain their polarization throughout the trajectory. Assume now that the processes of slowing down, stopping, and waiting do not depolarize the muons.[18] In this case, the electrons emitted from the target may [viz., if parity is not conserved in the decay process producing them] have an angular asymmetry about the polarization direction, e.g., for spin $\frac{1}{2}$ of the form $1 + a\cos\theta$. In the absence of any vertical magnetic field, the counter system will sample this distribution at $\theta = 100°$. We now apply a small vertical field in the magnetically shielded[19] enclosure about the target, which causes the muons to precess at a rate of $(\mu/s\hbar)H$ radians per sec. The probability distribution in angle is carried around with the μ-spin. In this manner we can, with a fixed counter system, sample the entire distribution by plotting counts as a function of magnetizing current for a given time delay." The point here is that during the fixed time-delay interval, the muon spin precesses in the horizontal plane by an amount proportional to the magnetic field. It therefore changes the angle between the spin direction and the direction of motion of the detected electron, which is the angle whose distribution is sought. "A typical run is shown in Fig. [10-5]. As an example of a systematic check, we have reduced the absorber in the telescope to 5 in. so that the end-of-range of the main pion beam occurred at the carbon target. The electron rate rose accordingly by a factor of 10, indicating that now electrons were arising from muons isotropically emitted by pions at rest in the target. No variation in counting rate with magnetizing current was observed. . . . The only conceivable effect of the magnetizing current is the precession of the muon spins."

The authors deduced "as necessary consequences of these observations" the high degree of polarization of the muon beam and the marked anisotropy of the decay electron distribution; these facts, in turn, showed that parity was not conserved in the two decay processes of the π-μ-e chain. The value $\frac{1}{3}$ for the asymmetry parameter a was deduced from the ratio of peak to valley counting rates in Fig.

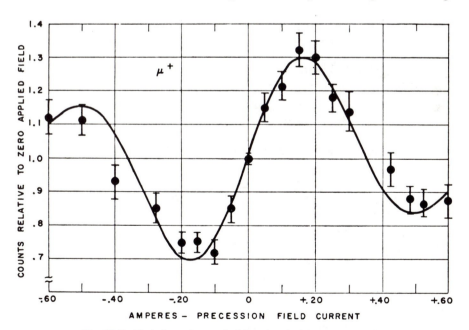

Fig. 10-5. Variation of overall delayed coincidence rate in the Nevis experiment with current in the coil surrounding the target. The solid curve is a fitted curve used in estimating the coefficient of $\cos\theta$ in the angular distribution. [*Phys. Rev.* **105** (1957), p. 1416, Fig. 2.]

10-5, and confirmed by the good fit to the points of the curve calculated using that value.

The same reaction had been under study for some time at the University of Chicago by a postdoctoral researcher, Jerome I. Friedman, and his research adviser, V. L. Telegdi. As they noted, "It is easy to observe the pertinent correlation by bringing π^+ mesons to rest in a nuclear emulsion[20] in which the μ^+ meson also stops. One has only to bear in mind two facts: (1) even weak magnetic fields, such as the fringing field of a cyclotron, can obliterate a real effect, as the precession frequency of a Dirac μ meson is $(2.8/207) \times 10^6$ sec^{-1}/gauss" so that, unless the line of motion of the muon were along the direction of the field, the original polarization would be destroyed; "(2) μ^+ can form 'muonium,' i.e., [the bound system] (μ^+e^-), and the formation of this atom can be an additional source of depolarization, both through its internal hyperfine splitting and the precession of its total magnetic moment around the external field.

"With these facts in mind, we exposed . . . nuclear emulsion pellicles (1 mm thick) to a π^+ beam of the University of Chicago synchrocyclotron. The pellicles were contained inside three concentric tubular magnetic shields and subject to $\leqslant 4 \times 10^{-3}$ gauss." The procedures are described somewhat more fully in a subsequent Article in *The Physical Review* (which also contained some additional data):

"Stacks of 1000μ . . . unsupported emulsion were exposed to a 40-MeV π^+ beam. . . . The emulsion received about 1.5×10^4 pions/cm^2 which is considered an optimum exposure as it produces a high density of events without obscuration.

". . . [T]he emulsion was shielded during the exposure from stray magnetic fields such as occur near a cyclotron. This was accomplished by placing the emulsion within the innermost of three concentric tubular magnetic shields. Each shield consisted of two layers: the outer layer was a medium permeability, high saturation steel, and the inner layer was a high permeability alloy. The magnetic field in the experimental area was of the order of 10 gauss; however, the emulsion in the shields was subject to less than 4×10^{-3} gauss, as measured with a calibrated flip coil.

"To insure most uniform development, the emulsion was processed unsupported and later glued to glass. Special attention was given to producing a reasonably high density for minimum ionization tracks . . . with a low background grain density in order to facilitate the locating of electrons.

"Events were detected by area-scanning under oil with a total magnification of 330. . . .

"In order to insure that the detection efficiency be completely independent of the direction of the positron from the decay of the muon, the observers were instructed to measure only those events that were detected by first seeing the π-μ decay. All events so found were used in the analysis except those in which the muon decays occurred within 50μ of either surface.

"When a π-μ decay was found, the scanner followed the muon to the end of its range and searched for the positron. A considerable effort was made to detect all positrons because it was felt that the loss of positrons could be a source of bias. In only about 1% of muon decays could a positron not be found. For each positron found, a check was made to insure that it indeed originated at the end of the muon track.

"For each π-μ-e decay found, the space angle θ between the initial direction of motion of the muon and that of the positron was measured. This was done to eliminate the effects of multiple scattering which change the momentum but not the spin orientation of the muon. The angles were measured with an accuracy of about $\pm 2°$. . . .

"2000 complete π-μ-e decays were analyzed . . . and the space angle θ defined above was calculated for each. For 60% of all events, chosen at random, this angle was recalculated independently; no appreciable discrepancies were revealed. From these data we find . . .

$$\epsilon = (B - F)/(B + F) = \ldots = 0.091 \pm 0.022."$$

Here B is the number of backward decays, i.e., with θ between $90°$ and $180°$, while F is the number of forward decays, with θ between $0°$ and $90°$. Evidently, there is "a preference for backward emission with respect to the direction of motion of the muon." An estimate of "errors of a systematic nature, such as those introduced by possible local variations in shrinkage [of the emulsion during development], etc.," showed that they are "small and of such a nature as to only decrease the experimentally observed value of ϵ." The numbers reveal a serious disadvantage of the emulsion method. In order to keep the emulsions from being swamped with tracks, the exposure must be kept fairly low; there are then relatively few events, with correspondingly large statistical uncertainties. Thus Friedman and Telegdi's Letter had reported the analysis of the first 1300 events, yielding a value for ϵ of 0.062 ± 0.027, somewhat smaller than the later value though not inconsistent with it; but the earlier value was less convincing because it differed from zero (the value it would have if parity were conserved) by less than three standard deviations.

In contrast to Garwin, Lederman, and Weinrich, who had assumed the form $1 + a\cos\theta$ for the distribution of electron directions, Friedman and Telegdi used the data to establish it: They analyzed the experimental distribution function $W(\theta)$ as a sum of suitable polynomials in $\cos\theta$ and found that "the best fit is obtained with a constant and a linear term only. . . ."

"With this information at hand, we determined $W(\theta)$ by a least-squares fit to a linear form in $\cos\theta$." The data and the best-fitting linear form are shown in Fig. 10-6. "This yields

$$W(\theta) = 1 - (0.174 \pm 0.038) \cos\theta."$$

The statistical analysis suggested that the indicated error was a true standard deviation, in which case the equation "implies that the asymmetry observed is real to an extremely high confidence level."

"As an additional check of possible bias, an analogous determination of the distribution of the muon direction of emission with respect to the direction of incidence of the pion beam was carried out. This distribution was found to be isotropic, within statistics, as would be expected." Specifically, the asymmetry parameter ϵ was found to be -0.026 ± 0.029.

Friedman and Telegdi noted that since the detailed nature of the two interactions involved was not known, there was no theoretical value of the coefficient a to be used as comparison; instead, the experimental value might, in principle, be used at a later date as a

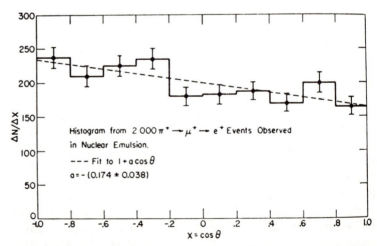

Fig. 10-6. Histogram of angular distribution of positrons from muon decay in the University of Chicago experiment (solid line), together with the best linear fit (dashed line). The error bars on the histogram indicate statistical standard errors. [*Phys. Rev.* **106** (1957), p. 1292, Fig. 1.]

test of theoretical proposals.[21] "Unfortunately, the experimental value of *a* is only indirectly related to the asymmetry coefficient predicted by theory. Muons can be depolarized by various causes during their lifetime, and the observed value of *a* represents only a lower limit." In fact, the Nevis experiment had established that the asymmetry measured in graphite was about twice that measured in nuclear emulsion, showing that there are some depolarization mechanisms that depend on the medium. In addition, as suggested earlier, if some of the muons formed muonium, and even if no other effects were operative, this posed special problems; and subsequent experiments at the University of Chicago, mentioned in a Note Added in Proof to the Article, had indicated that some muonium indeed was formed.

"Muonium will have a $^1S_0$ ground state and $^3S_1$ first excited state, separated by the hyperfine splitting [$\hbar \times (3 \times 10^{10}$ sec^{-1})] from it. If the μ^+ is initially in a completely polarized state, say $\alpha(\mu)$, while the electrons of the medium are, of course, unpolarized [$\alpha(e)$ and $\beta(e)$], two 'muonium' states can be formed, $\alpha(\mu)\alpha(e)$ and $\alpha(\mu)\beta(e)$. The first of these is an exact eigenstate of muonium ($^3S_1$, $m = 1$), while the latter is not, corresponding to a coherent superposition of $^1S_0$ and $^3S_1$ states with $m = 0$. [Because of the nature of the capture process,] $\alpha(\mu)\alpha(e)$ and $\alpha(\mu)\beta(e)$ will be formed in equal amounts, leading to a net time-dependent value of the muon spin z component after capture given by $\langle S_z(t) \rangle = \frac{1}{4}(1 + \cos\omega t)$, where $\omega \cong 3 \times 10^{10}$ sec^{-1}. Thus 50% of the muons captured into muonium will be effectively depolarized in a time short compared to their mean life for decay (2×10^{-6} sec). Hence if x is the fraction of muons forming muonium, a theoretical asymmetry coefficient a will be reduced to $a(1 - x/2)$, if no further depolarization mechanisms are at work.

"In a given external magnetic field, $^3S_1$ muonium would precess at an about one hundred times faster rate (1.4×10^6 sec^{-1}/gauss) than the free muon. This was the reason for our elaborate shielding precautions.

"Conversely some depolarization mechanisms too slow to affect appreciably the spin of a free positive muon during its lifetime could relax the total spin of $^3S_1$ muonium quite effectively during the same time. If such mechanisms were in action, an asymmetry coefficient $a(1 - x)$ would be observed even in perfectly shielded emulsion. . . . [A] succession of several 'charge exchange' processes would lead to about the same reduction in asymmetry.

"In comparing our results with those obtained in experiments in which the muon spin is made to precess by the application of an external field, the preceding remarks have to be borne in mind."

Nevertheless, since even in the possible presence of such effects the asymmetry was observed to be nonzero, "one can conclude that both interactions responsible for the transitions in the decay chain $\pi^+ \to \mu^+ \to e^+$ are not invariant under . . . space inversion."

The news created considerable excitement, not only among physicists but among laymen as well; *Time* magazine, for example, carried a five-column story about it. There was one further group to be heard from. A group at Brookhaven National Laboratory, headed by R. Ronald Rau, felt that "it seemed desirable to investigate the π^+ - μ^+ - e^+ angular distribution integrated over the whole momentum spectrum for μ^+ mesons stopped in liquid hydrogen." Knowing (as they did when they submitted their publication) of the still unpublished results of the other two groups that had studied this reaction, they reasoned as follows: "It is likely that the most effective depolarizing effect on the stopping μ^+ meson results from the capture and loss of electrons in the last fraction of its range.[22] It is then plausible that the depolarization of the slowing μ^+ mesons might be least in substances with high ionization potentials and there may be less depolarization in hydrogen than in carbon." Their results were also published as a Letter in *The Physical Review*, but later than any of the others.

This experiment is by far the simplest of the four to describe. "One-hundred-Mev π^+ mesons, produced by bombarding a copper target with one-Bev protons in the Cosmotron,[23] were selected by magnetic analysis and directed into a liquid hydrogen bubble chamber, 6 in. long, 2 in. deep, and 3 in. high. Appropriate absorber was placed in the beam so that the π mesons stopped in the chamber. According to Lee and Yang, the μ mesons are polarized with their spin along the direction of emission from the stopped π mesons. Since the π mesons decay isotropically, the μ mesons are polarized at random with respect to any magnetic field in the chamber and their precession in the field will tend to destroy angular correlations. Therefore a degaussing coil was placed around the chamber which reduced the field in the chamber to less than 0.25 gauss." Presumably the tedious job of scanning the pictures and measuring the appropriate events was accomplished by conventional methods, as no mention is made of any special techniques or precautions.

"Figure [10-7, part (a)] shows the μ - e angular distributions derived from 980 events. A similar plot of the π - μ angles shown in Fig. [10-7, part (b)] is consistent with the spherical symmetry to be expected from this decay. Since the μ range of about 1.1 cm is small compared with the dimensions of the chamber, and since the μ directions are isotropic, it seems unlikely that any scanning or measuring bias effects the measured distributions appreciably." The distribution

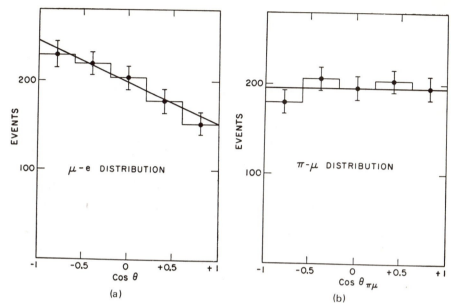

Fig. 10-7. (a) The angular distribution of the electrons from μ^+ decay in the Brookhaven bubble chamber, together with a least-squares fit. [*Phys. Rev.* **105** (1957), p. 1928, Fig. 1.] (b) The angular distribution of the muons from π^+ decay in the Brookhaven experiment, together with a least-squares fit which gives $1 + (0.043 \pm 0.045) \cos\theta$. [Ibid., p. 1928, Fig. 2.]

was again of the form $1 - a\cos\theta$, with $a = 0.25 \pm 0.045$. It was further confirmation, though somewhat anticlimactic.

There is a peculiar postscript to this chapter. To begin with, Lee and Yang in their first paper on parity nonconservation did not mention one of the consequences most easily susceptible of test: that electrons produced through β decay should be longitudinally polarized (here, again, the lack of reflection invariance is manifested in the presence of a screw motion). They did, however, discuss it early in 1957 in a paper suggesting a theoretical construct that would provide an explanation in the β-decay and π - μ - e decay cases.[24] Their prediction was checked by a group at the University of Illinois, under Hans Frauenfelder; the work was published in *The Physical Review* in May 1957. Approximately two years later, Lee Grodzins of Brookhaven National Laboratory pointed out, in the *Proceedings of the National Academy of Sciences*, that the same sort of study had been made nearly thirty years before: "[I]n 1928, R. T. Cox, C. G. McIlwraith, and B. Kurrelmeyer carried out an investigation of double scattering of beta rays from radium which can now be interpreted as . . . evidence for parity violation in weak interactions."

The geometrical considerations in such a study are shown in Fig. 10-8. Part *A* shows the situation for an initially unpolarized beam. Here both targets are thin; the first scattering serves merely to polarize the beam, while the second scattering provides the analysis. The polarization is transverse and in such a direction that the second scattering gives greater intensity when the source is at the $180°$ position than when it is at $0°$ but no difference between the $90°$ and $270°$ positions. Part *B* shows the case of a beam initially polarized longitudinally, illustrated for the particular case of the spin and momentum in opposite directions. Here again it is the second scattering, from a thin target, that acts as an analyzer; the first scattering in this case is from a thick target, giving multiple scattering which converts the polarization from longitudinal to transverse. In this case, the significant asymmetry is between the $90°$ and $270°$ positions of the source.

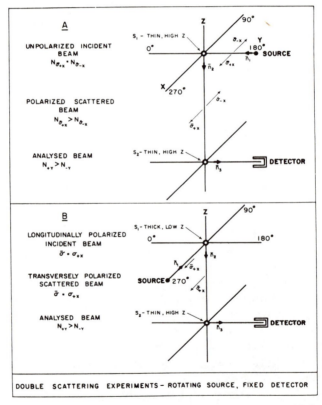

Fig. 10-8. Geometry for double scattering. *A*, for an initially unpolarized beam; *B*, for an initially longitudinally polarized beam. [*Proc. Natl. Acad. Sci. (U.S.)* **45** (1959), p. 400, Fig. 1.]

Cox, McIlwraith, and Kurrelmeyer observed just such a $90°$-$270°$ asymmetry.[25] Their apparatus, especially the counter, was not wholly trustworthy, and the theory relating the scattering asymmetries to polarization was not yet published; nevertheless, they were convinced that the effect was real, and they suggested that "the source of the asymmetry must be looked for . . . in some asymmetry of the electron itself." The work was extended by Carl T. Chase, a student of Cox, who improved the apparatus in a number of ways and obtained results which, while smaller in magnitude, still confirmed the qualitative aspect of the original work. Meanwhile, also, the theory of polarization in double scattering had been published. But there were still enough loopholes in the experiment that its implications were not accepted.

What makes the situation so peculiar is that, as Grodzins realized even before his paper appeared, the asymmetry reported by Cox et al. was of the wrong sign. Grodzins was sufficiently intrigued so that he and a student, Sidney Altman, duplicated the original apparatus (except that instead of radium, they used ^{90}Sr as a source). In a series of runs with various targets, they found an asymmetry of the expected sign, opposite to that found by Cox et al. Grodzins has given his opinion that, though Cox denies it, "there was an unfortunate error at the time which crept in between the data taking and the paper writing."

The point is somewhat academic, of course, because the early results were not accepted. But in 1957 there was no room for doubt.

Notes

1. This is quoted from *Physics for Poets*, by Robert H. March, p. 31. Copyright 1970, by McGraw-Hill, Inc. Used with permission of McGraw-Hill Book Company.

2. This term refers to the property of rotating the plane of polarization of light passing through a solution or liquid phase of the substance.

3. In fact, the synthesis of such a substance from optically inactive components yields a mixture of equal quantities of the two forms.

4. This is the form in which the rotation is to the left.

5. So were two other laws that had become significant only with the advent of quantum mechanics and which, though not properly conservation laws, are somewhat related to the law of conservation of parity. One is the law of invariance under charge conservation, which states that to every state of a system of particles there corresponds another possible state identical with the first in every respect except for the replacement of each particle by its antiparticle. The other is the law of invariance under time reversal, which states that for every possible state of a system, there is another possible state identical with the

first except for the reversal of direction of every velocity and every intrinsic spin, with the particles tracing backwards the paths they traced in the original state. Both of these also hold in classical mechanics, provided the quantum mechanical term *antiparticle* is replaced by appropriate classical phraseology.

6. The differing modes of decay imply different interactions with other particles, which would be expected to produce a difference in the masses.

7. The τ and the θ were among a group of particles that were originally referred to as "strange" because of an apparent discrepancy between the copiousness with which they were produced, implying a strong interaction, and the comparative slowness with which they decayed, indicating only a weak interaction. The resolution of this discrepancy, which is discussed in detail in Chap. 15, involved the introduction of a new quantum number called "strangeness," which can have only integral values. For the present purposes, it is sufficient to note that nucleons and pions have strangeness 0; the τ^+ and the θ^+ have strangeness +1.

8. See Chaps. 7 and 8.

9. An assembly of particles with intrinsic angular momentum is said to be polarized if the spins are predominantly parallel to some specified direction.

10. The effect is an indirect one: the field acts on the atomic electrons, which in turn influence the nuclei through the hyperfine interaction.

11. For a paramagnetic substance, the magnetic intensity H and the magnetization M constitute a pair of thermodynamic variables parallel to pressure P and volume V of an ideal gas. Thus, just as adiabatic expansion lowers the temperature of a gas, so adiabatic demagnetization lowers the temperature of a paramagnetic salt.

12. The photograph in Fig. 10-2 shows two.

13. Each counting rate is plotted as of the end of the counting interval.

14. In fact, the experiment also shows that charge-conjugation invariance does not hold; the arguments will not be given here. As far as can be seen, however, the combination of coordinate reversal and charge conjugation is a valid symmetry of nature.

15. Note that this implies that the muons have a sort of a screw motion; the screw is, of course, one of the most obvious instances of lack of reflection invariance in ordinary experience.

16. At the time, s was known only to be half an odd integer. A value of $\frac{1}{2}$ was considered most likely, but $\frac{3}{2}$ was a strong possibility. In fact, one of the results of the experiment was additional support for the value $\frac{1}{2}$.

17. A widespread story, possibly apocryphal, says that when Garwin first joined the group, plans were under way actually to rotate a counter system around the target, and Garwin said, in effect, "Why don't you rotate the spins instead?"

18. This assumption is not strictly valid. Cf. the discussion of the question by Friedman and Telegdi, below.

19. It is shielded against the fringing field of the cyclotron magnet, which was usually about 20 gauss and pointed upward.

20. This is photographic emulsion in which the concentration of the heavy elements (silver and bromine), with which the charged particles interact to produce developable grains, is especially high.

21. Weinrich, in his thesis, compared the results of the Nevis experiment with various theoretical proposals.

22. Cf. the remarks of Friedman and Telegdi, p. 171.

23. "Cosmotron" was the name applied to Brookhaven's proton synchrotron.

24. But it did not provide an explanation for the original τ-θ puzzle, as the new proposal dealt with the nature of the neutrino, which is involved in the π and μ decays and in β decay but not in the pionic decay modes of the τ-θ.

25. Their second target was thick, not thin, but a measurable effect should still have been present.

Bibliography

The first three Letters: C. S. Wu, E. Ambler, R. W. Hayward, D. D. Hoppes, and R. P. Hudson, *The Physical Review* **105**, 1413 (1957); Richard L. Garwin, Leon M. Lederman, and Marcel Weinrich, *The Physical Review* **105**, 1415 (1957); Jerome I. Friedman and V. L. Telegdi, *The Physical Review* **105**, 1681 (1957). A follow-up to the last Letter appeared in Jerome I. Friedman and V. L. Telegdi, *The Physical Review* **106**, 1290 (1957). The details of the Nevis experiment are contained in Marcel Weinrich, Ph.D. thesis, Columbia University, 1958 (unpublished; issued as Report No. NEVIS-56, under a joint ONR-AEC Program). The Brookhaven work appears in A. Abashian, R. K. Adair, R. Cool, A. Erwin, J. Kopp, L. Leipuner, T. W. Morris, D. C. Rahm, R. R. Rau, A. M. Thorndike, W. L. Whittemore, and W. J. Willis, *The Physical Review* **105**, 1927 (1957).

For a more extensive discussion of the roles of Lee and Yang, see Jeremy Bernstein, "Profiles: A Question of Parity," *New Yorker Magazine*, 12 May 1962, p. 49.

The work on β-ray polarization was reported in H. Frauenfelder et al., *The Physical Review* **106**, 386 (1957). The early double-scattering work is treated in L. Grodzins, *Proceedings of the National Academy of Science* **45**, 399 (1959).

See also Philip Morrison, *Scientific American* **157**, No. 4, 45 (1957); C. N. Yang, in *Nobel Lectures*, vol. III, p. 393, and T. D. Lee, ibid., p. 406.

The three original Letters and Grodzins's paper, together with background material and commentaries by the authors, are reproduced in *Adventures in Experimental Physics*, vol. γ (World Science Education, Princeton, New Jersey, 1974), p. 93.

Chapter 11

Recoilless Emission and Absorption of Radiation

Most experiments of outstanding significance are so because of the influence they have within their own field, either in providing a deeper insight or in opening a new path. Occasionally, however, an experiment achieves landmark status chiefly because of its effect on other fields. This chapter describes one such experiment: an effect discovered in the course of an investigation in nuclear physics, which has been of profound importance not only in nuclear physics (and there in ways only indirectly related to the original study) but in other fields as well—gravitation and solid state, to name some cases. As a result the discoverer, Rudolf Mössbauer of the Technical Institute of Munich and the Max Planck Institute for Medical Research in Heidelberg, was awarded a Nobel prize for physics in 1961.

Mössbauer was concerned with the interaction of the atomic nucleus with electromagnetic radiation, the expectation being that this would provide clues to the dynamical behavior of nucleons, just as the study of electromagnetic properties of atoms had led some fifty years earlier to a theory of the dynamcis of atomic structure. In the case of the atom, one useful tool had been resonance fluorescence, in which the atom is excited by absorbing light of one of its own natural frequencies and then reradiates. This phenomenon had not been readily usable for nuclei, because of the recoil effect, described by Mössbauer as follows:

"Nuclear resonance fluorescence of γ radiation can be observed only with difficulty, because the γ quanta on their emission and absorption suffer so large a recoil energy loss, as a consequence of giving up recoil momentum to the emitting and absorbing nuclei, that the emission and absorption lines are substantially shifted relative to one another and thus the resonance condition is violated."

To put this on a more quantitative basis, consider the emission of a photon of energy $E_\nu = h\nu$ by a system of mass M, initially at rest. Denote the transition energy, that is, the difference in energy between the initial and final states of the resting system, by E_0. On emission, this energy is shared between the photon and the recoiling system. It is easy to calculate the relationship between E_ν and E_0. The photon has momentum E_ν/c; the recoiling system has an equal and opposite momentum and therefore a kinetic energy $(E_\nu/c)^2/2M = E_\nu^2/2Mc^2$. The sum of this and E_ν is the total transition energy:

$$E_\nu + E_\nu^2/2Mc^2 = E_0 . \tag{11-1}$$

Now Mc^2 is the rest energy of the emitting system, which is of the order of billions of electron volts for either an atom or a nucleus. On the other hand, the transition energy E_0, and with it the photon energy E_ν, are at most a few million electron volts, even for nuclei, and only a few to a few thousand electron volts for atoms. Consequently, to a good approximation, the solution to Eq. (11-1) is[1]

$$E_\nu \cong E_0 - E_0^2/2Mc^2 .$$

A parallel argument shows that the energy E available for excitation of a system of mass M by absorption of a photon of energy E_ν is approximately

$$E \cong E_\nu - E_\nu^2/2Mc^2 .$$

It is to be noted that a given E_ν is less than the E_0 that gave rise to it, and greater than the E that it can excite; the two are separated by the order of E_0^2/Mc^2.

For atomic transitions, this difference is only a few millionths of E_0, and is comparable with—often less than—the natural width of spectral lines. For nuclei, however, the ratio of the difference to the transition energy is of the order of a few thousandths, while natural linewidths may be as small as a few parts in 10^{13} of the transition energy. Thus while in the atomic case there is still significant overlap between the energy distribution of the emitted photons and that of the absorbing atomic states, this is not true for nuclear radiation.

It is obvious that the recoil effect could be effectively eliminated if the mass M were made larger by many orders of magnitude. Mössbauer's contribution was to show that this is indeed possible: that under appropriate circumstances, when nuclei are bound in a crystal lattice they do not behave as if they were free, but rather the recoil momentum from a radiative transition may be taken up by the crystal as a whole. This means that M is the mass of the crystal, some 10^{23} times that of the nucleus.

Actually, other methods had been devised for circumventing the difficulty; Mössbauer made his initial discovery in the course of an experiment utilizing one of them, which was published in 1958 in *Zeitschrift für Physik*. All such methods involved the production of a Doppler shift of the gamma ray by putting the emitting and/or absorbing nuclei into motion by some mechanism: the use of an ultracentrifuge, the recoil motion resulting from a previous emission or absorption process, or thermal motion at a relatively high temperature. Mössbauer chose the last of these. However, all previous experiments had been, according to Mössbauer, "in the form of scattering studies, in which always the quanta resonantly scattered by the nuclei must be sorted out from the background of elastic scattering and the radiation scattered by the Compton effect." This led to considerable experimental difficulty, as well as to rather severe restrictions on the nuclides to which the methods could be applied.

Mössbauer observed, "The difficulties mentioned can be avoided if the nuclear resonance effect is measured in absorption. Nevertheless, since the effect, especially for soft gamma radiation, is very small compared with the absorption effect of the atomic shell, extreme demands are placed on the precision and stability of the measuring apparatus for the measurement of the lifetime[2] of a nuclear level by an absorption experiment. . . .

"Nuclear resonance fluorescence has particular interest in the energy range of soft gamma rays, since for low temperatures the influences of chemical binding in solids are to be expected in this energy range.

"In the present work the lifetime τ_γ of the 129-keV level in Ir^{191} was determined. Studies at the temperature of liquid O_2 showed a strong influence of chemical binding on the interaction cross section for nuclear absorption. . . . The binding effect is very sensitively dependent on the vibration spectrum of the solid."

For reasons that will be explained later, the experiment had to be done by a comparison method. "Figure [11-1] shows the experimental arrangement, Fig. [11-2] the construction of the absorber-cryostat. The absorbers, two rolled sheets, one of iridium and one of

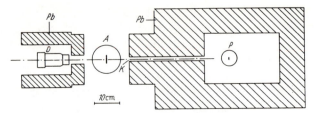

Fig. 11-1. Schematic arrangement of Mössbauer's first experiment. *A* is the absorber cryostat, *P* the cryostat containing the source, *D* the detector (a sodium iodide scintillator crystal and a photomultiplier), and *K* the collimator. *A* and *P* were set on a heavy stand. [*Z. Phys.* **151** (1958), p. 132, Fig. 2.]

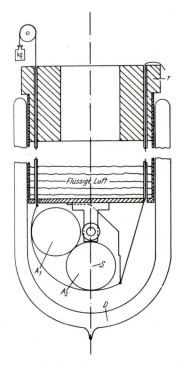

Fig. 11-2. Schematic diagram of the absorber cryostat, showing the arrangement by which either of two absorbers A_1 and A_2 could be brought into the beam incident at *S*, by pulling a cord. *D* is the wall of the Dewar flask, and *T* an insulating plug. *Flüssige Luft* means liquid air. [*Z. Phys.* **151** (1958), p. 133, fig. 3.]

platinum, each about 0.4 mm thick, 35 mm in diameter, were mounted in such a way that unrestrained contraction on cooling was possible.

"What was studied was the absorption of the 129-keV gamma radiation in iridium emitted upon β decay of Os¹⁹¹. .... [T]he observed spectrum . . . , in addition to the 16-d activity of Os^{191}, contains also components of the 95-d activity of Os^{195} . . . [including] hard lines of Re^{185} at 640 and 875 keV emitted on K capture[3] of Os^{185} . . . .

"While the self-absorption of the resonance line in the source increases strongly with the thickness of the preparation, the hard component of radiation from Re^{185} undergoes only an insignificant self-absorption. The radiation intensity emitted by the source thus is displaced toward the harder radiation components with increasing thickness of the slab of material. The substance was therefore arranged as a layer (area 80 mm^2) and the quantity limited to 0.6 g. . . . The . . . source was melted onto the bottom of a cylindrical quartz receptacle, serving for the receiving of liquid air, which with the preparation was in a Dewar vessel. The mounting of the source as described was necessary to keep to a minimum the motion of the substance as a consequence of contraction of the supporting arrangement on cooling. Such motion of the substance could lead to a change in the mean layer thickness of the absorber 'seen' by the source, if the former is not completely plane parallel. The chief experimental difficulty for the lifetime measurement lay in the sure elimination of such an influence on the measurement of a change in geometry from cooling. The problem was satisfactorily solved by the method described for supporting the source, by the use of approximately plane parallel absorbers of large area, and by the choice of a relatively large separation (at least 50 cm) of the source from the absorbers. . . .

"A direct determination of the interaction cross section for nuclear resonance absorption by a measurement of the total extinction coefficient is generally not possible, since the nuclear resonance absorption is ordinarily very small compared to the absorption effect of the atomic shells. The nuclear resonance effect in Ir^{191} was therefore determined by a difference measurement, in which the absorption of the resonance line in the resonant absorber iridium was measured at different temperatures under conditions such that the resulting intensity changes stood in direct relation to the cross section for resonance absorption and every side effect was eliminated. The cross section . . . is . . . a function of the temperatures of the source and the absorber. Because of the temperature dependence of the total absorption, all measurements were carried out at constant absorber temperatures, that is, in every instance only the temperature of the source was varied. The temperature dependence of the total absorption has its evident basis in the fact that the number of atoms per cm^2 of surface of an absorbing substance changes with temperature. This effect in the present case would have counterbalanced the nuclear resonance effect and completely masked it, which was avoided by restricting the temperature variations to the source. Of course, the self-absorption in the source also shows a change (smaller, to be sure) with temperature. This effect, which likewise counteracts the resonance effect, could however be eliminated experimentally by alternating intensity measurements with the resonant absorber and a comparison absorber. Since the self-absorption effect in the source affects the intensity behind both absorbers, while the nuclear resonance effect appears only for the resonant absorber, the first effect drops out of a difference measurement if the absorbers are so chosen that they absorb almost equally strongly. . . .

"Iridium ($Z = 77$) served as resonant absorber, platinum ($Z = 78$) as comparison absorber. The difference in intensity of the radiation penetrating the two absorbers amounted at room temperature to about 0.1%. What was measured was the total radiation intensities I_t^{Ir} and I_t^{Pt} behind the resonance absorber (Ir) and the comparison absorber (Pt). . . .

"For the measurements for the investigation of the effects of chemical binding on nuclear resonance absorption, the absorbers were always at the temperature of liquid O_2. The temperature of the source was varied between the boiling point of O_2 and that of water. . . ."

The directly determined quantity was the fractional difference $(I_t^{Ir} - I_t^{Pt})/I_t^{Pt}$, for various source temperatures. By an analysis that will not be reproduced here, this difference could be converted into a cross section $\bar{\sigma}_{ra}$ for resonant absorption.

"Figure [11-3] contains the results of the measurements in which the absorbers were cooled.[4] In [part (a)] are presented the differences of the intensities measured behind the resonant absorber and the comparison absorber, in [part (b)] the cross section . . . calculated therefrom. . . . [Part (b)] contains in addition to the experimental points the theoretical variation of the cross section . . . for two different frequency distributions of the vibrational spectrum of iridium." The theoretical analysis was adapted from a twenty-year-old treatment by Willis E. Lamb, Jr., of the resonance capture of neutrons in a crystal.

As can be seen, the two theoretical curves agree only qualitatively with the experimental results; but the lack of exact agreement was unimportant, as there were assumptions in the theory that were subject to modification (the solid and dashed curves show how a change in one of the assumptions influences the outcome). What was important was that both theory and experiment showed a strong increase in the cross section as the temperature of the source decreased. This is in sharp contrast to the intuitive argument that lower temperature would simply decrease the Doppler broadening of the line arising from thermal motion and thus decrease the amount of overlap.

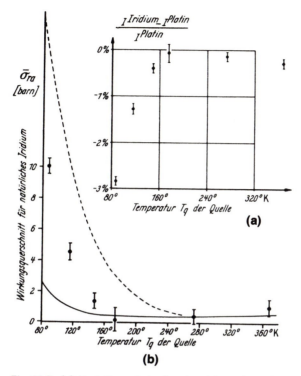

Fig. 11-3. (a) Variation of the fractional intensity difference in γ radiation after absorption in iridium and in platinum, as a function of the temperature of the source. The absorbers were at a temperature of 88 K. (b) Cross section for resonance absorption of γ rays in Ir^{191} as a function of temperature, derived from the data shown in part (a). [*Z. Phys.* **151** (1958), p. 140, Fig. 8.]

Mössbauer, in an Article in *Zeitschrift für Naturforschung* describing the further development of the concept, summarized the theory in the following way:

"The emission or absorption of a quantum by a nucleus bound in a crystal leads in general to an alteration of the vibrational state of the crystal lattice, which takes up the recoil momentum. On account of the quantization of the internal energy, the crystal can absorb the recoil energy only in discrete amounts. With decreasing temperature the probability of excitation of the internal states decreases more and more, so that for soft γ rays for a part of the quantum transitions the crystal as a whole takes up the recoil momentum. The quanta emitted or absorbed in this situation undergo practically no energy loss because of the large mass of the crystal, and ideally satisfy the resonance condition.

"Figure [11-4] shows the theoretical emission and absorption spectra of the 129-keV transition in Ir^{191} at a temperature of $88°$ K.

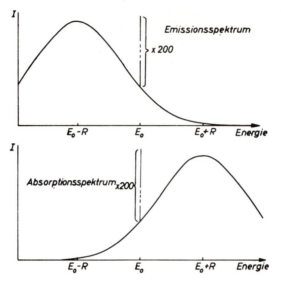

Fig. 11-4. "Position and shape of the emission and absorption spectra of the 129-keV transition in Ir^{191} at 88°K for a lifetime $\tau = 1.4 \times 10^{-10}$ sec. Zero point of the energy scale suppressed; units of ordinate arbitrary; height of the line at $E = E_0$ (resonance energy) shown reduced in the ratio 1:200." R is the recoil energy of a free nucleus, $E_0^2/2Mc^2$. [*Z. Naturforsch.* **14a** (1959), p. 213, Fig. 1.]

"The spectra each consist of two parts:

"(1) A broad distribution reflecting the thermal motion of the atoms bound in the crystal lattice. The quantum transitions falling in the range of this 'thermal' line are coupled to a change of vibrational state of the crystal lattice.

"(2) An extremely strong line with the natural linewidth, which contains those quantum transitions for which no energy loss appears, since the crystal as a whole takes up the recoil momentum. This 'recoilfree line' thus appears in emission and absorption unshifted at the position of the resonance energy E_0."

As the temperature decreases, the relative proportion of the second component to the first increases; it is this that causes the rise in the resonance-absorption cross section.

If the unshifted line was indeed as narrow as the natural linewidth, it ought to be possible to shift it off resonance by the Doppler effect of a rather small velocity, thereby both confirming the effect and obtaining a direct measurement of the width. Mössbauer proceeded to do so, and reported the results in a Letter in *Die Naturwissenschaften* and in the Article in *Zeitschrift für Naturforschung* already quoted. The following is also from the Article.

"In the present work using Ir[191] as an example the 'recoilfree' sharp emission and absorption lines were demonstrated by means of a 'centrifuge' method. In this the source was moved relative to the absorber, whereby through the Doppler effect the emission line was shifted to larger or smaller energy. By this shift of the emission line the complete overlap of the 'recoilfree' emission absorption lines, present for the source at rest, was removed. Consequently the resonance condition was violated and the strong resonance absorption effect of the 'recoilfree' line was made to disappear. An analysis of the 129-keV γ radiation penetrating the resonance absorber (iridium) as a function of the relative velocity of source and absorber then yields immediately the width of the 'recoilfree' line, that is, the natural linewidth, and thus also the lifetime of the 129-keV level of Ir[191] ....

"Figure [11-5] shows the experimental arrangement. The structure of the cryostats was described earlier.[5] The resonant absorber (iridium) and a comparison absorber (platinum) could be brought alternately into the radiation beam. The absorbers and the source were at the temperature of liquid O_2. The scintillation spectrometer was gated by a photocell in such a way that only those quanta were registered that were emitted by the source during its residence on the marked portion of its rotational path.

"Figure [11-6] shows the results of the measurements. Each individual datum point was determined from some 10 measurements of the radiation intensities behind each of the two absorbers. The total time for measurement amounted to 14 days. The radiation intensity behind the comparison absorber (platinum) was, within the limits of

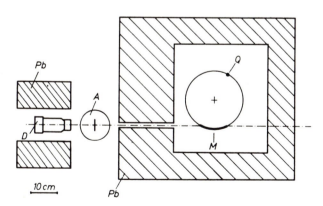

Fig. 11-5. "Experimental geometry. *A*, absorber cryostat; *Q*, rotating cryostat with source; *D*, scintillation detector. *M* is the part of the rotational path of the source used for the measurement." [*Z. Naturforsch.* **14a** (1959), p. 215, Fig. 2.]

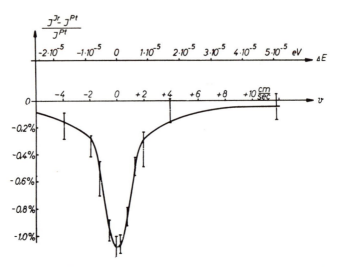

Fig. 11-6. Fractional intensity difference behind platinum and iridium absorbers, $(I^{Ir} - I^{Pt})/I^{Pt}$, as a function of the relative velocity of source and absorber. The upper scale shows the corresponding energy shift. [*Z. Naturforsch.* **14a** (1959), p. 215, Fig. 3.]

measurement accuracy, independent of the relative velocity v. The *mean* errors indicated were determined from the fluctuations of the individual measurements and are always larger than the statistical errors. ... The indicated curve through the measured values was calculated numerically ... and corresponds to a level width $\Gamma = (4.6 \pm 0.6) \times 10^{-6}$ eV for the 129-keV level in Ir^{191} ."

Mössbauer presented the method as a new tool for investigating very short-lived, low excited states; and in fact it yielded a value for the lifetime of the level involved that is considerably more reliable than any previously obtained. What turned out to be the most significant feature, however, was that the resonance condition could be satisfied with the source and absorber relatively at rest, whereas moderately small relative velocities of source and absorber corresponded to shifts away from the resonance condition that are large compared with the natural linewidth. Conversely, then, any external effect that would modify the original energy could be compensated for by imposing a relative motion; and the amount of modification could be determined by the value of the relative velocity needed to restore the resonance. It is this aspect that has made possible the subsequent developments ranging far outside the field of nuclear physics. The applications are numerous, and only a few can be described here.

An early application made use of the hyperfine structure of a nuclear transition to determine the magnetic moment of an excited nuclear level. The structure results from the splitting of the level (and of the ground-state level) by interaction of the magnetic moment with the internal magnetic field in a crystal; the intensity pattern depends on whether the magnetization in source and absorber are parallel or perpendicular to each other, and the spacings between the components depends in a simple way on the gyromagnetic ratios of the two states. A further refinement of this technique can be used to measure the coupling between the electric quadrupole moment of the state and the gradient of the electric field at the nucleus, since the coupling shifts the energies of different magnetic sublevels by different amounts and thus modifies the basic structure of the line.

Another application in nuclear physics involves the *isomer shift*. This is a change in the transition energy resulting from the electrostatic interaction of the nucleus with its own inner-shell electrons; it reflects the fact that the nucleus is not a point charge, and measurements of it provide information about the difference between nuclear radii in the ground and excited states.

There is also an effect on the energy called the *chemical shift*, again arising from the electron-nucleus interaction but referring to differences from one chemical environment to another. This evidently reflects the influence of the chemical state on the electron density "at" the nucleus—properly, averaged in some suitable way over the nuclear volume—and thus gives information regarding such things as the characteristics of chemical bonds. The most important applications in solid-state physics, however, have been the investigation of magnetic fields and electric field gradients by the use of isotopes with known magnetic dipole and electric quadrupole moments.

A particularly fascinating application is in a test of general relativity. The Mössbauer effect was used to measure the gravitational red shift produced by a difference in height of 74 feet, representing a shift in frequency of about five parts in 10^{15} and agreeing excellently with the value predicted by theory.

Other applications range from measuring an index of refraction for gamma rays to indicating the chemical states of iron in various ancient pigments and clays (which helps deduce the techniques used in the manufacture of the pottery). The Mössbauer effect is truly a versatile probe.

Notes

1. An easy way to see this is to solve the equation by successive approximations: The first approximation is to neglect the term in E_γ^2, giving $E_\gamma \cong E_0$; this approximation is then used for E_γ in the term in E_γ^2 to produce an improved

approximation, which is the one quoted. The process could be repeated if higher accuracy were desired.

2. The resonant absorption cross section depends on the natural linewidth Γ, which in turn is related to the lifetime by the uncertainty principle: $\tau = h/\Gamma$.

3. This is a form of radioactivity in which a nucleus of atomic number Z absorbs one of its own atomic electrons, usually from the K or innermost shell, and emits a neutrino, transforming into a nucleus of atomic number $Z-1$. It can take place, in principle, whenever positron emission can; and it may be possible even when positron emission is forbidden by energy restrictions.

4. For the lifetime measurements themselves, the absorbers had been kept at room temperature.

5. See p. 181.

Bibliography

The original papers are in German: R.L. Mössbauer, *Zeitschrift für Physik* **151**, 124 (1968); *Die Naturwissenschaften* **22**, 538 (1958); and *Zeitschrift für Naturforschung* **14a,** 211 (1959). English translations are presented in *The Mössbauer Effect*, edited by Hans Frauenfelder (W.A. Benjamin, Inc., New York, 1962).

The physics of the phenomenon is discussed in some detail in L. Eyges, *American Journal of Physics* **33,** 790 (1965).

See also Mössbauer's Nobel lecture, in *Nobel Lectures*, vol. III, p. 584, and *Mössbauer Effect: Selected Reprints* (American Institute of Physics, New York, 1963), which contains an annotated bibliography of further readings.

Chapter 12

Reality of the Neutrino

Most objects in nature, including subatomic entities, are detected first and then fitted into theoretical ideas. A few are first predicted theoretically, but the predictions are not taken too seriously until after they are confirmed by observation. The case of the neutrino is an outstanding exception. Its existence was postulated in 1930, and for the next 25 years its reality was accepted on faith—largely because the only apparent alternative was even more distasteful. Always, however, there was a wish that it could be independently observed. The fulfillment of this wish was finally reported in *Science* in 1956 by a group headed by Frederick Reines and Clyde L. Cowan, Jr.; a detailed account was published in *The Physical Review* in 1960. Their work is described in this chaper.

    The concept of the neutrino was introduced in order to account for a peculiarity in the phenomenon of β decay, which is the emission of an electron by a radioactive nucleus. It was learned quite early that the β rays emitted by a given element did not all have the same energy, or even a few discrete energies, but rather a continuous distribution such as shown in Fig. 12-1. This was very puzzling because a transition between two systems each with a definite, characteristic energy should involve the release of a similarly definite energy. Consideration of cases where a radioactive chain branched, with one element decaying sometimes by β emission and sometimes by α emission, and then rejoined, with two elements forming the same product, showed that the maximum or "end-point" energy was the value to be regarded as the proper difference. The question then was what be-

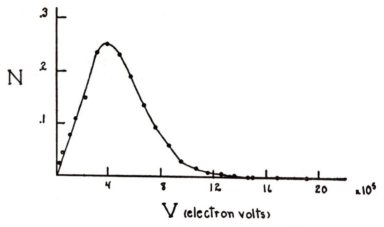

Fig. 12-1. The energy spectrum of β rays emitted by RaE (^{210}Bi). [*Phys. Rev.* 48 (1935), p. 394, Fig. 5.]

came of the excess when the β particle carried only part of that amount. One suggestion was that the β particle emerged from the nucleus with the full energy but lost part of it in an unspecified manner in passing through the outer portion of the atom; this, however, was disproved in 1927 by means of a calorimetric experiment by C. D. Ellis and W. A. Wooster, who found that the total energy carried by the β particles corresponded just to the average over the distribution rather than to the end point. The situation was further confounded when nuclear spins began to be determined, and it was discovered that there was also a discrepancy in the angular-momentum balance: The spins of the initial and final nuclei were always either both integer or both half-odd-integer multiples of $\hbar$, whereas the electron has a spin of $\frac{1}{2}\hbar$ and an orbital angular momentum that is an integer multiple of $\hbar$, and these values cannot be combined under the quantum mechanical rules of combining angular momenta in such a way as to balance.

One means of reconciling the discrepancy would be to give up the laws of conservation of energy and angular momentum, at least on the detailed microscopic level. No less authoritative a figure than Niels Bohr, in fact, maintained that in actuality the conservation laws held only on the average and not for every individual process. But Bohr's idea was, on the one hand, drastic enough to be unpalatable, and on the other, not drastic enough to rescue the situation and still fit the data on branched decays. Accordingly, in December 1930, Wolfgang Pauli made another suggestion. In a letter to some workers in radioactivity attending a meeting in Tübingen, he wrote,[1] "... I have hit on a desperate remedy to save the laws of conservation. . . . This is the possibility that electrically neutral particles exist which I will call neutrons, which . . . have a spin $\frac{1}{2}$ and obey the exclusion principle. . . . The mass of the neutrons should be of the same order as those of the electrons. . . . The continuous beta spectrum would then be understandable if one assumes that during beta decay with each electron a neutron is emitted in such a way that the sum of the energies of neutron and electron is constant."

Pauli repeated his proposal in a talk given by invitation at a meeting of The American Physical Society in June 1931; but he never published it,[2] perhaps feeling that the idea should not be taken that seriously. But Enrico Fermi took up the suggestion,[3] and built it into a theory of the β-decay process that had a striking success in accounting for the shapes of β-ray energy distributions. Thus the neutrino proved itself indirectly.

Further indirect evidence began to pile up in the next few years as measurements began to be made of the recoil directions and speeds of residual nuclei formed in the β-decay process. But arguments for the existence of the neutrino on the basis of such experiments involved essentially circular reasoning. A way to observe the neutrino directly remained to be found.

The problem lies in the properties of the neutrino. Since it is uncharged, it does not produce ionization; even if it had a nonzero rest mass, gravitational effects are so weak as to be essentially useless for studying subatomic entities; and it is presumably stable, so that it could not be observed through decay products. In fact, it presumably takes part in no interaction other than that giving rise to β decay, and, consequently, it must be detected through a reaction intimately related to β decay.

A typical β-decay process is written in chemical form as

$$_{Z}X^{A} \rightarrow {}_{Z+1}X'^{A} + e^{-} + \bar{\nu}. \tag{12-1}$$

Here X and X' are two nuclides, characterized by atomic numbers Z and $Z + 1$, respectively, and having the same mass number A; e^{-} represents the emitted electron, and $\bar{\nu}$ the (anti)neutrino.[4] The nuclear mass of $_{Z}X^{A}$ must be larger than that of $_{Z+1}X'^{A}$ by more than the mass of the electron in order for the reaction to proceed spontaneously. Now it is a general principle of elementary-particle physics that if a reaction proceeds in one direction, nothing other than conservation laws will keep it from proceeding in the other.[5] In particular, given the existence of process (12-1), the process

$$\bar{\nu} \; + \; _{Z+1}X'^{A} \; \rightarrow \; _{Z}X^{A} \; + \; e^{+}, \tag{12-2}$$

where e^{+} is the antielectron, or positron, is perfectly conceivable; moreover, if the antineutrino provides the energy equivalent to the surplus mass of the products over the reactants, the reaction can be expected to take place.

However, the interaction giving rise to the processes (12-1) and (12-2) is very weak. A convenient way to describe the strength of an interaction is by means of a dimensionless constant involving the characteristic coupling. For electromagnetism, the coupling is the fundamental unit of charge, e, and the dimensionless constant is $e^{2}/\hbar c = 1/137$. For the β-decay interaction, the corresponding dimensionless constant is about 10^{-10}.[6] Consequently, a neutrino must pass through many *light-years* of ordinary solid matter before the chances of its having triggered a reaction of the form of (12-2) reach one in two.

But the number of interactions can be increased *either* by increasing the number of targets—several light-years of solid—*or* by increasing the number of projectiles: instead of dealing with only a few neutrinos, using astronomical numbers of them. This became a meaningful suggestion, though, only after World War II. While the sun is a copious source of neutrinos,[7] it is of constant intensity; whereas what is needed is a source that can be turned on and off, if not at will, then anyway at known times, so that the difference (which is what is meaningful) can be observed.[8] The advent of large-scale nuclear fission devices was just what was needed, as each fission event gives rise to an average of about six β-decay processes, each producing one antineutrino.

The reaction chosen for study was the induced positron emission of a proton,

$$\bar{\nu} \; + \; p \; \rightarrow \; e^{+} \; + \; n. \tag{12-3}$$

This has a variety of advantages, the most obvious being that it is extremely easy to provide target protons in water. A second is that both of the product particles are readily observable by means of very characteristic processes that they undergo: for the neutron, capture by a nucleus with emission of γ rays, and for the positron, annihilation. A third is that the energy that must be supplied by the neutrino is not very large, so that significant fractions of the neutrinos produced in other reactions can induce this one.

At first, it was proposed to detect only the positron. In order to get a signal large enough to be meaningful against a large background of other processes that might simulate the desired reaction, this required a flux of neutrinos of a magnitude available only within a few hundred feet of a nuclear explosion. Fantastic as it may sound, plans were actually developed and test drillings begun for a scheme described by Cowan several years later as follows:

"We would dig a shaft near 'ground zero' about 10 feet in diameter and about 150 feet deep. We would put a tank, 10 feet in diameter and 75 feet long on end at the bottom of the shaft. We would then suspend our detector from the top of the tank, along with its recording apparatus, and back-fill the shaft above the tank.

"As the time for the explosion approached, we would start vacuum pumps and evacuate the tank as highly as possible. Then, when the countdown reached 'zero,' we would break the suspension with a small explosive, allowing the detector to fall freely in the vacuum. For about 2 seconds, the falling detector would be seeing antineutrinos and recording the pulses from them while the earth shock [from the blast] passed harmlessly by, rattling the tank mightily but not disturbing our falling detector. When all was relatively quiet, the detector would reach the bottom of the tank, landing on a thick pile of foam rubber and feathers (fig. [12-2]).

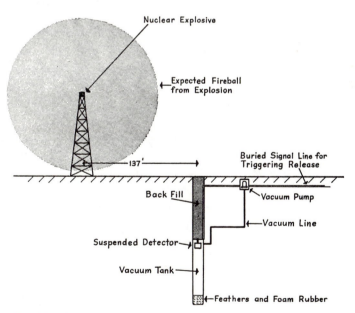

Fig. 12-2. Schematic arrangement of the proposed experiment to detect antineutrinos produced in a nuclear explosion. [*Smithsonian Report for 1964* (Smithsonian Institution, Washington, D. C., 1965), p. 419, Fig. 4.]

"We would return to the site of the shaft in a few days (when the surface radioactivity had died away sufficiently) and dig down to the tank, recover the detector, and learn the truth about neutrinos!"

This audacious plan received encouragement from a number of people, not the least of whom was Fermi. Fermi's support was significant because, in addition to being a competent theorist, he was a very astute experimenter. It was helpful to know, first, that he felt the importance of doing a neutrino-detection experiment at all, and second, that he regarded the proposed method as reasonable.

Before the experiment was set up, however, Reines and Cowan restudied the situation and recognized that if they made use of the neutron as well as the neutrino, they could utilize a much smaller flux of neutrinos, such as are available in the vicinity of a nuclear reactor. The final experiment was carried out at the Savannah River Plant of the U.S. Atomic Energy Commission.

"The detection scheme is shown schematically in Fig. [12-3]. An antineutrino ($\bar{\nu}$) from the fission products in a powerful production reactor is incident on a water target in which $CdCl_2$ has been dissolved.[9] By reaction [(12-3)], the incident $\bar{\nu}$ produces a positron (β^+) and a neutron (n). The positron slows down and annihilates with

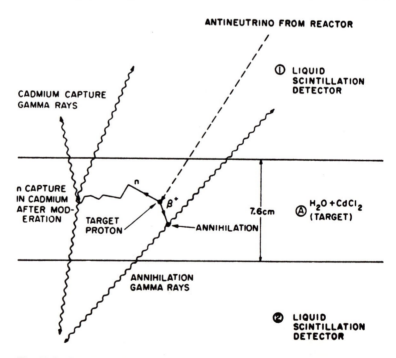

Fig. 12-3. Schematic diagram of the process by which the inverse beta process was detected. [*Phys. Rev.* **117** (1960), p. 159, Fig. 1.]

an electron in a time short compared with the 0.2-μsec resolving time characteristic of our system, and the resulting two 0.5-Mev annihilation gamma rays penetrate the target and are detected in prompt coincidence by the two large scintillation detectors placed on opposite sides of the target.[10] The neutron is moderated [that is, slowed down to thermal energy] by the water and then captured by cadmium in a time dependent on the cadmium concentration (in our experiments practically all neutrons are captured within 10 μsec of their production). The multiple cadmium-capture gammas are detected in prompt coincidence by the two scintillation detectors, yielding a characteristic delayed-coincidence count with the preceding β^+ gammas. The experiment consisted in showing that:

"(1) Reactor-associated delayed coincidences of the kind described above were observable at a rate consistent with the calculated $\bar{\nu}$ flux and the detector efficiency. . . .

"(2) The first prompt-coincidence pulse of the delayed-coincidence pair was due to positron-annihilation.

"(3) The second prompt-coincidence pulse of the delayed-coincidence pair was due to cadmium capture of a neutron.

"(4) The signal was a function of the number of target protons.

"(5) The reactor-associated signal was not caused by gamma rays or neutrons from the reactor.

"Throughout the experiment an effort was made to provide redundant checks of these several points. . . .

"A consideration of the cross section for reaction [(12-3)] averaged over the fission antineutrino spectrum ($\sim 10^{-43}$ cm^2) and the available $\bar{\nu}$ flux ($\sim 10^{13}$ cm^{-2} sec^{-1}) made it apparent that large numbers of target protons would be required. These were provided by two plastic target tanks containing 200 liters of water each, shaped as slabs 7.6 cm deep and 132 cm by 183 cm in lateral dimensions [see Fig. 12-4]. Each water tank was sandwiched between two of the three large liquid scintillation detectors (Fig. [12-5]) [see also Fig. 12-6]. The thickness of the water tanks was limited by the absorption of the 0.5-Mev positron-annihilation radiation produced in the antineutrino reaction. The array of tanks formed two 'triads' with one detector tank in common. The 58-cm depth of the scintillation detectors was chosen so as to absorb the cadmium-capture gammas with the maximum efficiency attainable in the space available for the system. Consideration of light-collection efficiency and the energy resolution required of the system resulted in the use of an extremely transparent liquid scintillation solution. . . .

"The tank walls were painted white, and each tank had 110 5-in. . . . photomultipliers (55 on each end) for collection of the scintillation light. . . . Figure [12-5] shows a sketch of the detectors

Fig. 12-4. One of the target tanks, in a version used for calibration and testing (hence the phototube bases projecting from the end). [*Smithsonian Report for 1964* (Smithsonian Institution, Washington, D. C., 1965), Plate 2, Fig. 1.]

set into their lead shield. The lead side walls and floor were 10.2 cm thick, the roof 20.3 cm thick, and the doors 20.3 cm (toward the reactor) and 15.2 cm (away from the reactor). The 5400-liter capacity of the detectors required scintillation-liquid storage tanks and pumping facilities at the reactor site [see Fig. 12-7].[11] Stainless steel pipes transferred the liquids some 75 meters from the outside storage tanks to the detector inside the reactor building. The detector was filled after emplacement in the shield.

"The [photomultiplier] tubes were checked individually prior to insertion in the tank. Noisy ones were rejected, and the gains were balanced by means of a resistor in the high-voltage circuit at each tube. . . . The 110 tubes of each tank were operated in parallel.

"Scintillation light in the detector was translated into electrical impulses by the photomultipliers on each tank, passed to preamplifiers, and then sent via coaxial cables to the remainder of the electronics located in a trailer van outside the building." Such extensive equipment has long since been dwarfed by the paraphernalia of high-energy experiments. For its time, however, the setup was almost unparalleled.

"To illustrate how the system functioned, we trace through a

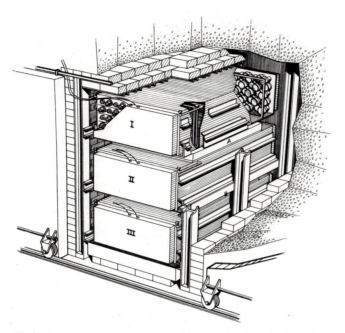

Fig. 12-5. "Sketch of detectors inside their lead shield. The detector tanks marked 1, 2, and 3 contained liquid scintillator solution which was viewed in each tank by 110 5-in. photomultiplier tubes. The white tanks contained the water-cadmium chloride target, and in this picture are some 28 cm deep. These were later replaced by 7.5-cm deep polystyrene tanks, and detectors 1 and 2 were lowered correspondingly." [*Phys. Rev.* **117** (1960), p. 160, Fig. 2.]

typical antineutrino-induced event: Let an event occur in the top target (*A*), producing a positron and a neutron by reaction ([12-3]). The sequence is analyzed by the equipment as follows: First, two pulses from the positron annihilation arise on the signal lines of counters 1 and 2, are amplified by the top triad amplifiers 1β and 2β, and are accepted by the 'top β^+ coincidence unit' if in the proper energy gates (0.2 to 0.6 Mev) and in time coincidence (<0.2 μsec apart). This unit informs the 'top n coincidence unit' that it has received a β^+-like pulse by sending a pulse which opens a gate in the n unit approximately 30 μsec long. The second pair of prompt coincidence pulses is accepted by the neutron-coincidence unit if they have the correct energies (>0.2 Mev in each tank with a total from 3 to 11 Mev . . .). If the second pulse occurs between 0.75 and 30 μsec after the first, the neutron unit announces the completion of the delayed-coincidence event by activating a scaler and triggering the sweeps of the recording oscilloscopes.

"During this process of selection the signals from the tanks have been stored on delay lines for presentation on the oscilloscope when

(a)

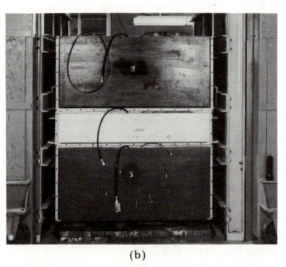

(b)

Fig. 12-6. (a) "A completed detector section ready for insertion in the shield. The tank is made of steel plate, with the exception of the bottom. This is a cellular aluminum structure, similar to aircraft skin sections, which provides strength against bending while affording little obstruction to the entry of gamma rays from below." [*Smithsonian Report for 1964* (Smithsonian Institution, Washington, D. C., 1965), Plate 2, Fig. 2.] (b) "The lower 'triad' of detectors used at the Savannah River Plant rests in its lead shield ready for test at the Los Alamos Scientific Laboratory. The dark rectangles labeled '1' and '2' are the ends of the large liquid scintillation tanks. . . . The target liquid is in the white center tank." [Op. cit., Plate 4, Fig. 1.]

Fig. 12-7. Liquid storage facilities. [*Smithsonian Report for 1964* (Smithsonian Institution, Washington, D. C., 1965), Plate 3, Fig. 1.]

the signal meets the acceptance requirements. . . . Film records of each oscilloscope trace were made on 35-mm . . . film. The camera-frame advance was controlled by the scope trigger. Typical antineutrino-induced events are shown in [Fig. 12-8, parts (a) and (b)]. Figure [12-8, part (c)] shows calibration frames, and [Fig. 12-8, parts (d) through (g)] are examples of rejected frames.

"The energy calibration of the system was obtained by means of the peaks in the energy spectrum resulting from the passage through the tanks of relativistic cosmic-ray μ mesons, 'meson-through peaks.' . . . By matching the pulse heights at which the through peaks were located with a precision pulser, energy bounds were set on discriminator circuits and film calibrations were made. The energy scales were set by multiplying the scintillator depth (58 cm) by the rate of meson energy loss (1.6 Mev/cm). . . .

"Through peaks were run at regular intervals and the system was checked and aligned at weekly intervals on the average. . . . The efficiency of the system for detecting positrons and neutrons was determined using positron and neutron sources."

The first step was "to demonstrate the existence of a $\bar{\nu}$-like signal and its dependence on reactor power, and hence on $\bar{\nu}$ flux." Several series of measurements were made for this purpose, comparing the number of "acceptable" events per hour while the reactor was operating to the rate of events when the reactor was shut down. Data from part of the first series are given in Table 12-I. As can be seen, there were a total of 1.63 events per hour when the reactor was

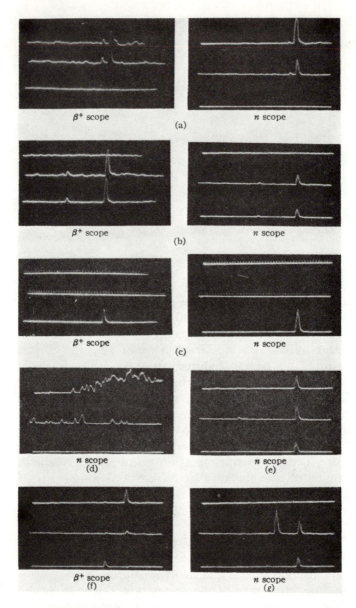

Fig. 12-8. Sample scope pictures. The three beams of each oscilloscope display respectively the outputs of the three detectors; the β^+ scope is operated at high gain, the n scope at low gain. (a) and (b) are acceptable frames, representing possible $\overline{\nu}$ events. (c) is a set of calibration frames. (d) through (g) are unacceptable events: (d) is the result of electrical noise, (e) and (f) cosmic-ray events, and (g) an event, possibly due to cosmic rays, rejected because of the appearance of an extra pulse (that in beam 2 of the n scope). [*Phys. Rev.* **117** (1960), p. 162, Fig. 5.]

Table 12-I. Partial data from the first series of measurements. [Adapted from *Phys. Rev.* **117** (1969), p. 163, Table I.]

Triad	$\bar{\nu}$ flux factor*	Running time (hr)	Total counts	Accidental counts†	Net rate (hr^{-1})
(a) Top	1.03	192.7	283	114	0.88 ± 0.10
Bottom	1.04	171.8	284	95	0.75 ± 0.10
(b) Top	0	67.3	55	31.8	0.34 ± 0.14
Bottom	0	69.7	44	39.7	0.06 ± 0.13

*A number proportional to the calculated flux of antineutrinos from the reactor.

†Calculated from the separate rates of single β^+ and n counts.

running and 0.40 when it was off, giving a net difference of 1.23 ± 0.24 events per hour associated with the operation of the reactor. "From these data we conclude that there was a reactor-associated signal." Subsequent series, made with increased cadmium concentration in the target and less stringent requirements on the pulse heights for the neutron counts, raised the rate to nearly 3 events per hour, thus increasing the confidence in the reality of the effect.

It remained, however, to establish that the effect was really what it was thought to be. To begin with, "it is important to establish that the counting rates observed were consistent with the expected cross section. For this purpose it is necessary to know the efficiency with which neutrons and positrons were detected.

"An attempt was made to determine the neutron-detection efficiency by studying the response of the apparatus to a Pu-Be neutron source.[12] Unfortunately, the neutrons from this source differed in two essential ways from the neutrons produced in our reaction:

"(1) The Pu-Be neutrons are of higher energy, ranging up to about 11 Mev; neutrons from our reaction are of the order of 10 kev;

"(2) The Pu-Be source is a point source, whereas the $\bar{\nu}$-associated neutrons are produced uniformly throughout the target volume.

"The second of these differences was eliminated in part by placing the source in several positions on top of the target tank and averaging the response. No simple, precise approach is available to take account of the differing spectra.

"First the relative neutron-detection efficiencies were measured for various source positions on the water target. . . . Then a measurement was made of the detection efficiency for a centrally placed source." The result was an overall value of 14% for the detection efficiency of $\bar{\nu}$-produced neutrons; this was taken as a lower limit, partly because of limitations imposed by the measurement technique and partly because of the difference in the neutron-energy spectrum. A calculated estimate of the probabilities that a neutron would not escape from the water target, that the capture process would take place during the time when the system was sensitive, and that the capture gamma rays would satisfy the coincidence criteria gave an efficiency of 24%. "It seems reasonable to state the efficiency as $\epsilon_n = 0.17 \pm 0.06$."

"The positron-detection efficiency was determined by dissolving a known amount of β^+ emitter, Cu^{64}, in the water targets and measuring the counting rate of prompt coincidences in the pair of detectors next to each target. . . .

"The calibration consisted of two parts: the measurement of the response to the Cu^{64} source and the determination of the source strength by calibration with a Na^{22} standard." The net result was a value $\epsilon_\beta = 0.15 \pm 0.02$.

"Using our experimental numbers, we are now in a position to calculate the cross section for the reaction $p(\bar{\nu}, \beta^+)n$ induced by antineutrinos from fission fragments. . . . [O]ur object is only to check whether the cross section which we deduce from our experiment is consistent with expectations. The cross section, σ, is calculated from the equation

$$\sigma = \frac{R}{3600\, F\, N\epsilon_n \epsilon_\beta} \ \ \mathrm{cm}^2,$$

where $R = 1.5 \pm 0.1$ hr^{-1}, the average signal rate per triad, $\epsilon_n = 0.17 \pm 0.06$, $\epsilon_\beta = 0.15 \pm 0.02$, $N = 1.1 \times 10^{28}$, the number of hydrogen nuclei in each target tank, and $F = 1.2 \times 10^{13}$ cm^{-2} sec^{-1}, the average $\bar{\nu}$ flux at the detector. Therefore . . .

$$\sigma = \left(1.2_{-0.4}^{+0.7}\right) \times 10^{-43} \ \mathrm{cm}^2.$$

This value is in agreement with the theoretically expected value of $(1.0 \pm 0.17) \times 10^{-43}$ cm^2."

This agreement alone might be taken as adequate evidence that the detectors were really registering what they were supposed to. But additional checks were needed to establish it beyond a doubt. The first test was to show that the first pulse had the proper characteristics to be ascribed to a positron. This was done in two ways. The first method was to examine the effects on the signal of various thicknesses of lead interposed between target tank B and detector 2. The results are shown in Table 12-II. However, this reduction in signal is a combination of two factors: It is "due primarily to the

Table 12-II. Signal rates for lead absorption experiment. [*Phys. Rev.* **117** (1960), p. 167, Table V.]

Pb thickness (cm)	Signal (hr^{-1})
0	1.24 ± 0.12
0.16	0.62 ± 0.14
0.48	0.40 ± 0.16
0.95	0.04 ± 0.17

Table 12-III. Absorption experiments with sources. [*Phys. Rev.* **117** (1960), p. 168, Table V; footnotes added.]

Pb thickness (cm)	Relative β^+ signal	Relative n signal	Relative $\bar{\nu}$ signal Predicted[*]	Relative $\bar{\nu}$ signal Observed[†]
0	1.00	1.00	1.00	1.00
0.16	0.47	(0.86)[‡]	0.40	0.50 ± 0.13
0.48	0.17	0.68	0.12	0.32 ± 0.14
0.95	0.04	0.45	0.02	0.03 ± 0.06

[*]Product of preceding two columns.
[†]Normalized values from Table 12-II.
[‡]Calculated value, not measured.

attenuation of the β^+ annihilation radiation and to a lesser extent to the attenuation of the gamma rays associated with neutron capture in cadmium." Each of these factors had to be determined in a separate mockup experiment: the first by "measuring the response of a NaI crystal placed above the middle of a water tank (containing a dissolved Cu^{64} source) as a function of the thickness of lead sheet interposed between tank and detector," and the second by measuring the "response of the lower triad to a neutron source placed at the top center of the water-$CdCl_2$ target . . . as a function of the lead thickness." The results are given and compared with the effect on the signal rate in Table 12-III. "The last two columns agree as well as might be expected in view of the uncertainties" arising from the difference between the mockups and the actualities.

The second method for the first test was to examine the pulse-height spectrum of the first pulses of $\bar{\nu}$-like events "to see if it was consistent with the spectrum to be expected from annihilation radiation." The spectrum is shown in part (a) of Fig. 12-9. "For comparison purposes, [part (b) of Fig. 12-9] shows two spectra, one

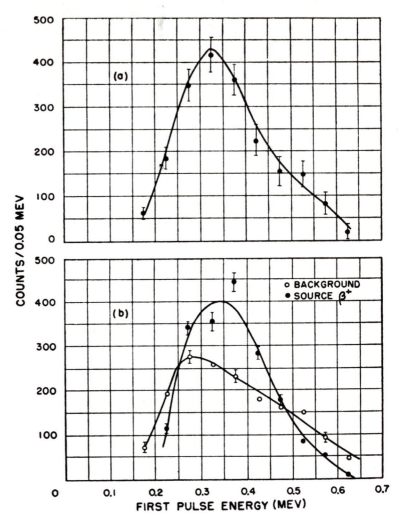

Fig. 12-9. "(a) Pulse-height spectrum, first pulses of $\bar{\nu}$-like events. (b) Background and β^+ source spectra for comparison purposes." [*Phys. Rev.* **117** (1960), p. 169, Fig. 13.]

from a Cu^{64} source, the other of background. . . . The spectrum of [part (a)] resembles that from the β^+ source more closely than that of the background shown in [part (b)]. This may be taken as evidence that the first pulses were due to positron annihilation."

Similarly, tests were made on the second pulse of each event to verify that it resulted from neutron capture in cadmium. "Evidence that the second pulses of $\bar{\nu}$-like events were due to neutrons is provided primarily by the shape of the time-delay spectrum and the effect on this spectrum of varying the cadmium concentration. Three separate runs were made, with cadmium-hydrogen atomic ratios of $\alpha = 0.005$, 0.010, and 0.

"... The time-delay distributions [for $\alpha = 0.005$ and 0.010] are in satisfactory agreement with the theoretical curves and the increase in cadmium concentration ... cause[s] a definite shift of the experimental distribution towards shorter capture times, as expected.

"In the case $\alpha = 0$ (top triad, no cadmium in target) for 118.4 hr of running with the reactor at the same power level as for the $\alpha = 0.010$ case, ... [there was obtained] a net rate of -0.02 ± 0.07 hr^{-1}. This value is consistent with a purely random or accidental background and is to be compared with the corresponding figure for the $\alpha = 0.01$ case: 1.92 ± 0.01 hr^{-1}."

For these pulses also, "the pulse-height spectrum ... was examined to see if it was consistent with the spectrum to be expected from neutron capture in cadmium." Again, the spectrum differed significantly from that of background, in the manner expected.

"In order to see whether the signal rate was proportional to the number of target protons, an experiment was performed in which 0.47 of the target protons were replaced by deuterons." Antineutrinos can interact with deuterons, giving the reaction

$$\bar{\nu} + d \rightarrow \beta^{+} + 2n;$$

but the cross section for this reaction is only 1/15 that for reaction (12-3). Accordingly, the event rate should be reduced by a factor

$$0.53 + [(2/15) \times 0.47] = 0.59$$

(the factor 2 arising from the production of two neutrons). At the same time, "the neutron-moderation properties of the medium were different, which resulted in a change in the capture time and also altered the leakage of the neutrons out of the target, and hence the capture efficiency." Corrections for these factors were estimated, and the expected counting rate was calculated as 0.84 ± 0.11 hr^{-1}. The actual net reactor-associated rate was measured as 0.58 ± 0.13 hr^{-1}, giving a ratio of observed to expected rates of 0.69 ± 0.019. The authors state, "Although the precision of the result leaves something to be desired, it has been shown that the reactor signal does depend on the presence of protons in the target."

Finally, it was necessary to assure that the signal was not due to other particles emanating from the reactor. "The only known particles other than antineutrinos which are capable of being produced by the fission process and a secondary reaction outside the detector are gamma rays, electrons, protons, and neutrons. A direct test of the reactor-associated signal is to measure how it is affected by neutron and gamma-ray shield. Protons and electrons are automatically excluded as the source of the signal if these more penetrating radiations can be ruled out. If the signal is due to antineutrinos it will, of course, not be affected by the shield. An experiment was done in which the correlated delayed coincidence rate was measured with and without a shield of 76 cm of wet sawdust in bags (density 0.52 g/cm^3).[13] The signal with the shield was measured with the top triad to be . . . 1.74 ± 0.12 hr^{-1} after adjusting it for the difference in $\bar{\nu}$ fluxes for identical runs without the shield. This number is to be compared with 1.69 ± 0.17 hr^{-1} for the runs with no shield."

The sawdust shield was placed at the front and back of the detector, these being "much less well shielded from the reactor" than the other faces; it overlapped "by a few feet the edges of the heavy shielding already there." Its thickness and density were such that it "should have reduced fast neutrons from the reactor by at least a factor of 10 and gamma rays by a factor of 5 or more." Its effectiveness "was checked by means of an Am-Be source placed in a standard position outside the lead shield door," and "reduced the count rate in the appropriate energy range by two orders of magnitude. The lack of appreciable change in the counting rate after installation of the sawdust shield is evidence that the signal was not due to neutrons or gamma rays produced outside the detector."

The evidence was firm. The elusive neutrino had finally been trapped, and there was proof that it "existed as an objective, demonstrable fact of nature."

Notes

1. Pauli himself, only thirty at the time, missed the meeting in order to attend a dance.

2. The first published mention of the idea was in a speech by S. Goudsmit at a meeting on nuclear physics held in Milan in October 1931 by the Royal Academy of Italy.

3. It was Fermi who suggested the name "neutrino," on the grounds that "neutron" had been introduced a decade earlier by W. D. Harkins for a neutral particle comparable in mass with the proton. Pauli's particle was much less massive, so that the (Italian) diminutive ending "-ino" was appropriate.

4. Originally, there was thought to be no significance to a distinction between neutrino and antineutrino. For basically aesthetic reasons, the choice used in Eq. (12-1) was made. This has turned out to be very convenient, as it permits the assignment in a natural way of a quantum number known as *lepton number*, which seems to be universally conserved.

5. According to Reines, this principle, known as *microscopic reversibility*, was one whose applicability the experimenters proposed to check.

6. Gravitation is even weaker, with a dimensionless constant of about 10^{-39}; nuclear forces, in contrast, are quite strong, with a constant of the order of unity.

7. It is actually an antineutrino that is needed for process (12-2); but neutrinos could drive another inverse of process (12-1), namely, $v + {}_Z X^A \rightarrow {}_{Z+1} X'^A + e^-$.

8. Here is where the need for many light-years becomes significant. The interposition of the earth or the moon is enough to block off not only the light from the sun, but the entire spectrum of electromagnetic radiation and practically all atomic or subatomic particles—except neutrinos!

9. In a first version of the experiment, done in 1953, the scintillator itself served as the target. That version gave a positive result agreeing with the theory; but it was inconclusive, largely as a result of a strong background caused by cosmic rays. The later version used a counting arrangement that allowed much better discrimination against this background; in addition, it was carried out underground rather than on the surface.

10. The use of such large liquid scintillators was, in itself, such a novel technique that the success of early tests of them was reported in a Letter in *The Physical Review*.

11. Even the storage tanks were specially designed. The three 1200-gallon tanks were "coated on their interior surfaces with an epoxy paint to preserve the purity of the liquids"; and, as the temperature of the scintillation liquid had to be maintained above $60°$F, they "were wrapped with several layers of insulating material on their outsides," and "long strips of electrical heating elements were embedded in the exterior insulation."

12. Since the early 1930s, conveniently compact neutron sources have been formed by intimately mixing beryllium with a long-lived α-particle emitter—in this case, plutonium. The α particles induce the nuclear reaction ${}_4Be^9 + {}_2He^4 \rightarrow {}_6C^{12} + {}_0n^1$, which gives neutrons with an energy of about 4.7 MeV greater than the energy of the incident α particle.

13. "In recognition of the Southern hospitality which . . . [the experimenters] were enjoying all this time," a pound of hominy grits was incorporated into the shield.

Bibliography

The complete report is F. Reines, C. L. Cowan, Jr., F. B. Harrison, A. D. McGuire, and H. W. Kruse, *The Physical Review* 117, 159 (1960).

A nontechnical account, including a discussion of much of the work that led up to the final experiment, is given by Clyde L. Cowan, Jr., in the *Smithsonian Report for 1964* (Smithsonian Institution, Washington, D.C., 1965), pp. 409-430. See also F. Reines and C. L. Cowan, Jr., *Physics Today* 8, No. 12 (1957).

Chapter 13

The Maser
and the Laser

Of the interactions that can take place between an atom and electromagnetic radiation, two are completely familiar. An atom in its ground state can absorb radiation of such a frequency as to raise it to one of its excited states; and an atom already in an excited state can spontaneously drop to a lower state by emitting radiation. The change in energy ΔE in both cases is, of course, related to the frequency ν of the corresponding radiation by the equation $\Delta E = h\nu$, with h being Planck's constant.

There is a third type of interaction which is much less well known: If an atom in an excited state is acted on by radiation of a frequency that it could naturally emit, it may produce *induced*, or *stimulated*, radiation. This process seems little different from spontaneous emission, but it is a separate, additional mode of interaction. Indeed, its existence was first recognized (by Albert Einstein in 1916) from the fact that it is needed to account for the equilibrium distribution of energy states of atoms in interaction with radiation.

For over 35 years thereafter, even though consideration of stimulated emission was essential in theoretical work, it had little or no place in the practical realm. For example, a classic monograph on the quantum theory of radiation published in 1954 gives the topic scarcely more than passing mention, while devoting considerable attention to such phenomena as resonance fluorescence (cf. Chap.

11) and Raman scattering.[1] There were occasional references in the literature to the "negative absorption" that would result from the process, but no thought of how it might be put to use.

With the development of microwave technology during World War II, the situation became ripe for a change. Much of this development had entailed at least some consideration of the interaction of microwaves with matter, especially with gases; and much of it had been carried out by people trained as physicists, who were thus prepared at the end of the war to probe more deeply into such interactions. Their work led to the growth of the field known as *microwave spectroscopy*, done initially at a few places (mostly industrial laboratories) and then spreading rapidly.

One of the first of these centers was Columbia University. There, as early as 1951, a group headed by C. H. Townes recognized that stimulated emission ought to be usable as the basis of a very high-resolution microwave detector, a microwave oscillator, or a microwave amplifier. By 1954, they had constructed and utilized such a device and verified its properties. Partly in recognition of the importance of this work, Townes was awarded a share of the 1964 Nobel prize in physics.[2] Further study led to extension of the principle to visible wavelengths; several approaches were taken at various places, the first success being achieved in 1960 by T. H. Maiman of Hughes Research Laboratories. The resulting devices were called the *maser*—an acronym for *m*icrowave *a*mplification by *s*timulated *e*mission of *r*adiation—and the *laser*—*l*ight instead of *m*icrowave. This chapter describes the work of Townes's group, as reported in a Letter to the Editor and two subsequent Articles in *The Physical Review*; and, after a sketch of the intervening developments, the work of Maiman, reported in a series of publications in *Physical Review Letters*, *Nature*, *British Communications and Electronics*, and *The Physical Review*.

The reason for the importance of these devices, and the basis for some of their special properties, is a characteristic feature of stimulated emission: The emitted wave has exactly the same direction *and phase* as the stimulating wave. It is useful to regard the mechanism by which either a maser or a laser operates as the triggering of transitions from excited states by photons already emitted in such triggered transitions, the series having been initiated by a photon of either thermal radiation or input radiation. Consequently, all the resulting photons are described by waves that are perfectly in phase. In contrast, the photons of spontaneous emission are randomly related in phase and direction. The high degree of coherence reveals itself in the time domain as a high degree of monochromaticity, and in the space domain as a large distance over which the phase of the wave is uniform. The latter feature is especially remarkable in the case of the laser and permits focusing of large amounts of optical power onto areas of the order of a wavelength square, thus giving electric field intensities of the order of megavolts per centimeter.

It is necessary first to examine the conditions under which stimulated emission can be observed. In matter at an absolute temperature T, the number n_i of atoms in a state of energy E_i is proportional to $\exp(-E_i/kT)$, where $k = 1.38 \times 10^{-16}$ erg/K is Boltzmann's constant. Between any two levels, say 1 and 2 with $E_2 > E_1$, the probability B_{12} for absorption per atom in state 1 is equal to the probability B_{21} for stimulated emission per atom. The *number* of absorption events is proportional to $n_1 B_{12}$, and the number of stimulated emission events to $n_2 B_{21}$. Since $B_{12} = B_{21}$, the relative numbers of events are governed by the exponential factors in the n_i's, which means that there is more absorption than stimulated emission. This is one reason why stimulated emission is of little direct importance in ordinary situations. In order to utilize the process, it is necessary to produce a distortion of the normal distribution of the atoms among the energy states—preferably, in fact, to produce a *population inversion*, an actual excess of atoms in the upper state over the number in the lower state.

Townes and his coworkers J. P. Gordon and H. J. Zeiger worked with the ammonia molecule, NH_3. In a classical picture, this is shaped like a triangular pyramid, with the three hydrogen atoms at the corners of the base and the nitrogen atom at the apex. When quantum mechanics is taken into consideration, the description must be somewhat modified to take account of the fact that there are two equivalent positions for the nitrogen atom, one on each side of the plane of the hydrogens. The potential energy of the nitrogen atom as a function of its distance from the plane of the hydrogens is shown qualitatively in Fig. 13-1. Here the dashed lines show what would be valid if only one side or the other of the plane were accessible. There would be two states for each allowed energy value, one for each side. In actuality, because there is only a finite potential "hump" between the two wells, the two states interact with each other to give two new states. In these states, the nitrogen atom cannot be said to be on one side or the other of the hydrogen plane; rather, it has equal probabilities of being on either side. The wave function describing one of the two states is unaltered by an interchange of the two positions, while the other wave function has its sign changed. The interaction splits the energies of the two new states,[3] the symmetric one being somewhat lower than the antisymmetric. The energy separation between members of a pair increases with increasing pair energy, but for the ammonia molecule it corresponds to frequencies in the microwave range.

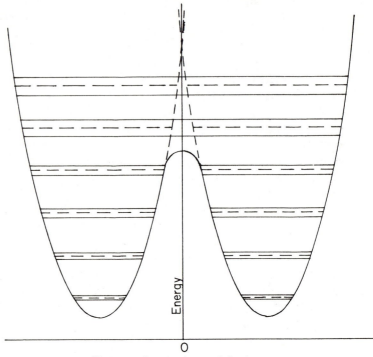

Fig. 13-1. Qualitative sketch of the potential energy of the nitrogen atom in the ammonia molecule, with representative energy levels. The solid lines denote the actual situation; the dashed lines, what it would be if only one side of the hydrogen plane were accessible.

In addition to these states, which are vibrational in character, the molecule also has rotational states, corresponding to rotations about either of two inequivalent axes—one perpendicular to the hydrogen plane, the other lying in it. The rotational states are identified by the values of the (quantized) angular momenta about each of these axes. The rotation alters the vibrational potential energy curves somewhat, by centrifugal stretching of the molecule, so that the separation of each vibrational pair depends on the rotational state. In the work of Gordon, Zeiger, and Townes, the transition principally used was that between the lowest vibrational pair in the rotational state having three units of angular momentum about each axis, called the *3-3 state*. This transition has a frequency of 23.870 MHz.

Another important property of the ammonia molecule is that, although it does not have a permanent electric dipole moment, an applied electric field will induce one, and its sign is opposite for the two members of a vibrational pair. If the field is inhomogeneous, there will be a force on the molecule,[4] which then also has opposite signs (i.e., directions) for the two members of a pair.

With this background, it is appropriate to turn to the description of the experimental work, drawn largely from the second of the two Articles.[5]

"The device utilizes a molecular beam in which molecules in the excited state of a microwave transition are selected. Interaction between these excited molecules and a microwave field produces additional radiation and hence amplification by stimulated emission. . . .

"A molecular beam of ammonia is produced by allowing ammonia molecules to diffuse out of a directional source consisting of many fine tubes. The beam then traverses a region in which a highly nonuniform electrostatic field forms a selective lens, focusing those molecules which are in upper inversion states while defocusing those in lower inversion states. The upper inversion state molecules enter a resonant cavity in which downward transitions to the lower inversion states are induced. A simplified block diagram of this apparatus is given in Fig. [13-2]. The source, focuser, and resonant cavity are all enclosed in a vacuum chamber.

"For operation of the maser as a spectrometer, power of varying frequency is introduced into the cavity from an external source. The molecular resonances are then observed as sharp

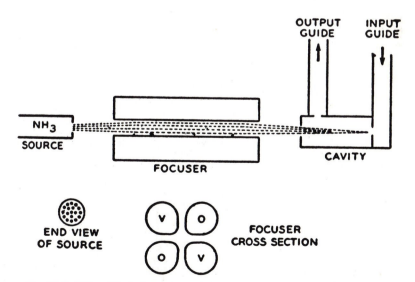

Fig. 13-2. "Simplified diagram of the essential parts of the maser." [*Phys. Rev.* **99** (1955), p. 1265, Fig. 1.]

increases in the power level in the cavity when the external oscillator frequency passes the molecular resonance frequencies.

"At the frequencies of the molecular transitions, the beam amplifies the power input to the cavity. Thus the maser may be used as a narrow-band amplifier. Since the molecules are uncharged, the usual shot noise[6] existing in an electronic amplifier is missing, and essentially no noise in addition to fundamental thermal noise is present in the amplifier.

"If the number of molecules in the beam is increased beyond a certain critical value the maser oscillates. At the critical beam strength a high microwave energy density can be maintained in the cavity by the beam alone since the power emitted from the beam compensates for the power lost to the cavity walls and coupled wave guides. This oscillation is shown both experimentally and theoretically to be extremely monochromatic.

"The geometrical details of the apparatus are not at all critical. . . . Two ammonia masers have been constructed with somewhat different focusers. Both have operated satisfactorily.

"A source designed to create a directional beam of the ammonia molecules was used. An array of fine tubes is produced . . . as follows. A 1/4-in. wide strip of 0.001-in. metal foil (stainless steel or nickel, for example) is corrugated by rolling it between two fine-toothed gears. This strip is laid beside a similar uncorrugated strip. The corrugations then form channels leading from one edge of the pair of strips to the other. Many such pairs can then be stacked together to form a two-dimensional array of channels, or, as was done in this work, one pair of strips can be rolled up on a thin spindle. The channels so produced were about 0.002 in. by 0.006 in. in cross section. The area covered by the array of channels was a circle of radius about 0.2 in., which was about equal to the opening into the focuser. Gas from a tank of anyhdrous ammonia was maintained behind this source at a pressure of a few millimeters of mercury. . . .

"The electrodes of the focuser were arranged as shown in Fig. [13-2]. High voltage is applied to the two electrodes marked V, while the other two are kept at ground. . . .

"In the first maser which was constructed the inner faces of the electrodes were shaped to form hyperbolas with 0.4-in. separating opposing electrodes. The distance of closest approach between opposing electrodes was 0.08 in., and the focuser was about 22 in. long. Voltages up to 15 kv could be applied to these electrodes before sparking occurred. In the second maser the electrodes were shaped in the same way, but were separated from each other by 0.16 in. This allowed voltages up to almost 30 kv to be applied, and somewhat more satisfactory operation was obtained since higher field gradients could be achieved in the region between the electrodes. This second focuser was only 8 in. long. Teflon spacers were used to keep the electrodes in place. To provide more adequate pumping of the large amount of ammonia released into the vacuum system from the source the focuser electrodes were hollow and were filled with liquid nitrogen.''

The operation of the focuser was analyzed in some detail in the first of the two Articles, written by Gordon alone.

''The field of the focuser exerts forces on the molecules due to their induced dipole moments. ... [The electrode] arrangement produces an electric potential in the gap of the form $\phi = \phi_0 r^2 \cos 2\theta$, where r and θ are cylindrical coordinates of a system whose axis is the axis of the focuser. The magnitude of the electric field produced by this potential is proportional to r. The force exerted on the molecules by this field is [proportional to the product of the field strength and its radial gradient, the first factor arising from the field dependence of the induced dipole moment and the second from the form of the force on a dipole]. It is radial and proportional to r for small fields strengths. The force is directed inward for molecules in upper inversion states ... and outward for molecules in lower inversion states.''

The remainder of the description is, again, from the second Article.

''The resonant cavities used in most of this work were circular in cross section, about 0.6 in. in diameter by 4.5 in. long, and were resonant in the TE_{011} mode[7] at the frequency of interest (about 24 kMc/sec). Each cavity could be tuned over a range of about 50 Mc/sec by means of a short section of enlarged diameter and variable length at one end. A hole 0.4 in. in diameter in the other end allowed the beam to enter. The beam traversed the length of the cavity. ... The diameter of the beam entrance hole was well beyond cutoff for this mode[8] and so very little loss of microwave power from it was encountered. ...

''Microwave power was coupled into and out from the cavities in several ways. Some cavities had separate input and output wave guides, power being coupled into the cavity through a two-hole input in the end of the cavity furthest from the source and coupled out through a hole in the sidewall of the cavity. In other cavities the sidewall hole served as both input and output, and the end-wall coupling was eliminated. ...

''Three ... diffusion pumps ... were used to maintain the necessary vacuum of less than 10^{-5} mm Hg. Nevertheless, due to the large volume of gas released into the system through the source, satisfactory operation has not yet been attained without cooling the focuser electrodes with liquid nitrogen. . . . [T]he cold electrode surfaces provide a large trapping area which helps maintain a low pressure in the vacuum chamber. . . . The solidified ammonia which builds up on the focuser electrodes is somewhat of a nuisance as electrostatic charges which distort the focusing field tend to build up on it, and crystals form which can eventually impede the flow of gas. For the relatively short runs, however, which are required for spectroscopic work, this arrangement has been fairly satisfactory.

"Experimental results have been obtained with the maser as a spectrometer and as an oscillator. Although it has been operated as an amplifier, there has as yet been no measurement of its characteristics in this role."

For use of the device as a spectrometer, the frequency of the input microwave power was varied; as it passed through the frequency of a molecular transition, the output power was increased by the power emitted as induced radiation from the molecules. Gordon's paper had reported the study of the ammonia spectrum by this method. "Resolution of about seven kc/sec was obtained, which is a considerable improvement over the limit of about 65 kc/sec imposed by Doppler broadening in the usual absorption-cell type of microwave spectrometer. . . . The sensitivity of the maser was considerably better than that of other spectrometers which have had comparably high resolution. . . .

"The experimental results obtained with the maser in its role as an oscillator agree with the theory . . . and show that its oscillation is indeed extremely monochromatic, in fact, more monochromatic than any other known source of waves. Oscillations have been produced at the frequencies of the 3-3 and 2-2 inversion lines of the ammonia spectrum. . . . Other ammonia transitions, or transitions of other molecules could, of course, be used to operate a maser oscillator.

"The frequency of the $N^{14}H_3$ 3-3 inversion transition is 23870 mc/sec. The maser oscillation at this frequency was sufficiently stable in an experimental test so that a clean audio-frequency beat note between the two masers could be obtained. This beat note, which was typically at about 30 cycles per second, appeared on an oscilloscope as a perfect sine wave, with no random phase variations observable above the noise in the detecting system. The power emitted from the beams during this test was not measured directly, but is estimated to be about 5×10^{-10} watt. . . .

"It was found that the frequency of oscillation of each maser could be varied one or two kc/sec on either side of the molecular transition frequency by varying the cavity resonance frequency about the transition frequency. If the cavity was detuned too far, the oscillation ceased. The ratio of the frequency shift of the oscillation to the frequency shift of the cavity was almost exactly equal to the ratio of the frequency width of the molecular response (that is, the line width of the molecular transition as seen by the maser spectrometer) to the frequency width of the cavity mode. This behavior is to be expected theoretically. . . . [T]he appearance of the beat indicates a spectral purity of each oscillator of at least . . . 4 parts in 10^{12} in a time of the order of a second.

"By using Invar cavities maintained in contact with ice water to control thermal shifts in their resonant frequencies, the oscillators were kept in operation for periods of an hour or so with maximum variations in the beat frequency of about 5 cps or 2 parts in 10^{10} and an average variation of about one part in 10^{10}. Even these small variations seem to be connected with temperature effects."

The narrow resolution of the maser as a spectrometer, its extreme monochromaticity as an oscillator, and its expected narrow bandwidth as an amplifier are all related to the fact that it operates as a regenerative amplifier, i.e., one in which part of the output is fed back in phase with the input. The bandwidth of any resonant circuit is determined by its quality factor Q, defined as the ratio of the maximum power stored in the circuit to the power dissipated during one cycle (for a simple RLC circuit, this leads to the relationship $Q = \omega L/R$); the bandwidth is roughly inversely proportional to Q. The positive feedback effectively reduces the net power dissipated, by replacing it with part of the output power; thus Q is increased, and the bandwidth is decreased. The effect is larger, the larger the power output of the oscillator or the gain of the amplifier. In the case of the maser, the linewidth can be reduced substantially below the natural width of the molecular transition for spontaneous emission.

This very small bandwidth, and the lack of tunability, form rather severe restrictions on the use of the ammonia maser as an amplifier. However, the amplifier also has a low noise figure,[9] and the authors suggested that it might be useful for certain special applications such as a "signal . . . from outer space, where the temperature is only a few degrees absolute." Such applications of the molecular beam maser have been of little or no significance, however: Once the utility of the principle had been demonstrated, it was only a short time before it was applied to other transitions, particularly that between two spin states of an electron in a paramagnetic solid where the frequency depends on the strength of the applied magnetic field.

Once the validity of the principle of the maser had been proven, it was natural to try to extend it to optical frequencies. The step, however, was far from trivial. Some of the problems were discussed, and solutions to them proposed, in a joint paper in *The Physical Review* in 1958 by Townes and Arthur L. Schawlow of the Bell Telephone Laboratories. Schawlow's training in optical spectroscopy stood him in good stead in dealing with the most easily recognized problem: The use of a resonant cavity, which worked extremely well for microwaves, simply could not be extended even to the infrared, let alone to the visible.

"To maintain a single isolated mode in a cavity at infrared frequencies, the linear dimension of the cavity would need to be of the order of one wavelength which, at least in the higher frequency part of the infrared spectrum [and *a fortiori* in the optical region], would be too small to be practical. Hence, one is led to consider cavities which are large compared to a wavelength, and which may support a large number of modes within the frequency range of interest. . . .

"Let us now examine the extent to which the normal line width of the emission spectrum of an atomic system will be narrowed by maser action, or hence how monochromatic the emission from an infrared or optical maser would be. . . . Assume for the moment that a single mode can be isolated. Spontaneous emission into this mode[10] . . . produces a finite frequency width . . .

$$\Delta\nu_{\text{osc}} = (4\pi h\nu/P)(\Delta\nu)^2,$$

where $\Delta\nu$ is the half-width of the resonance at half-maximum intensity, and P the power in the oscillating field. . . .

"For the case considered numerically [earlier in the paper], $4\pi h\nu\Delta\nu/P$ is near 10^{-6} when P is [equal to the minimum needed to sustain oscillation], so that $\Delta\nu_{\text{osc}} \sim 10^{-6}\Delta\nu$. This corresponds to a remarkably monochromatic emission. However, for a multimode cavity, this very monochromatic emission is superimposed on a background of stimulated [*sic*; 'spontaneous' apparently meant] emission which has width $\Delta\nu$, and which, for the power P assumed, is of intensity equal to that of the stimulated emission. . . .

"Another problem of masers using multimode cavities which is perhaps not fundamental, but may involve considerable practical difficulty, is the possibility of oscillations being set up first in one mode, then in another—or perhaps of continual change of modes which would represent many sudden jumps in frequency."

The problem, in other words, was to get the equivalent of a resonant cavity, in terms of selective response to a single oscillatory mode, without relying on dimensional consonance. Schawlow and

Townes proceeded to show that this could effectively be done by using a "cavity" with slightly transparent end walls, to provide coupling to the exterior, and no side walls—in short, just the configuration of a type of optical interferometer called *Fabry-Perot*. With this arrangement, all but one or a few desired modes would be lost through the sides before they could build up enough intensity to be troublesome.

The authors examined design parameters of a possible infrared device, using potassium vapor as the active substance and a potassium lamp as the exciter, and found them not unreasonable. They also considered the possibility of using other types of active materials; at least one, cesium vapor excited by a helium lamp, was eventually actually realized. They were somewhat less optimistic about optical maser action using solids as the active material,[11] partly because the spectral lines are generally broader, thus making mode selection more difficult, and partly because of the limited availability of pumping radiation of appropriate frequency. They recognized, however, that "there may be even more elegant solutions. Thus it may be possible to pump to a state above one which is metastable. Atoms will then decay to the metastable state (possibly by nonradiative processes involving the crystal lattice) and accumulate until there are enough for maser action."

Indeed, less than a year later, Schawlow was ready to suggest that ruby (aluminum oxide, Al_2O_3, with a fraction of a percent of chromium oxide as an "impurity") might be used as an active material: "There is a broad absorption band in the green and others in the ultraviolet. When excited through these bands, the crystal emits a number of sharp bands in the deep red (near 7000 Å). The two strongest lines (at 6919 Å and 6934 Å) go to the ground state, so that they will always have more atoms in their lower state, and are not suitable for laser action. However, the strongest satellite line (at 7009 Å)...goes to a lower state which is normally empty at liquid-helium temperatures, and might be usable."

In the same talk, Schawlow mentioned studies by Ali Javan of Harvard University on the transfer of energy in collisions between two kinds of gas in a mixture, which could lead to the necessary population inversion. This mechanism would eventually be utilized in an important class of lasers. But it was ruby that was to be the first successful medium, and that, not through the mechanism suggested in Schawlow's talk, but more nearly through the mechanism proposed by Schawlow and Townes.

Ruby, in addition to being an important gem stone, is a useful substance for the physicist. For one thing, it can be synthesized fairly easily. Moreover, the crystal structure is simple enough to be tractable but not so simple as to be trivial; in addition, the chromium ions have magnetic and optical properties that can be utilized readily. In fact, ruby was a widely used medium for solid-state masers. Maiman was one of a number of people working with ruby in this connection and thus was concerned with the behavior of the chromium ions. The first steps that led toward his development of the laser were the determination that when a ruby crystal is excited with green light, most of the return to the ground state takes place with the emission of the red fluorescence near 6900 Å, and the demonstration that significant depletion of the ground-state population could be produced. These developments were reported in *Physical Review Letters* in June 1960.

"The predominant processes which ensue in a fluorescent material when it is irradiated at an appropriate wavelength are shown in Fig. [13-3]. W_{13} is the induced transition probability per unit time due to an exciting radiation and the S_{mn} are decay rates which include both radiative and nonradiative processes. In this crystal S_{21} is easily obtained from the decay rate of the fluorescent level (^2E) after an exciting source is turned off. The lifetime for this process is

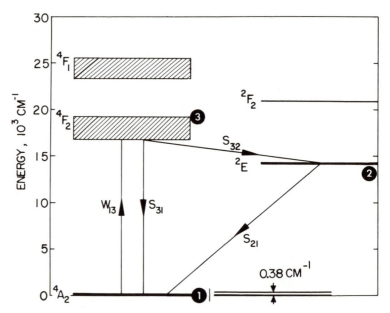

Fig. 13-3. "Pertinent features of the energy level diagram in ruby." The shaded areas represent broad bands of energy. The symbols at the level positions are a system of describing certain symmetry properties of the levels. [*Phys. Rev. Lett.* **4** (1960), p. 564, Fig. 1.]

about 5 msec. Varsanyi, Wood, and Schawlow have further demonstrated that this lifetime is almost entirely due to spontaneous emission, i.e., S_{21} is approximately the Einstein coefficient A_{21}.

"An approximate value for the rate S_{32} was obtained in the following way. A crystal of ruby was irradiated with 5600A radiation causing absorption into the lower band ($^4A_2 \rightarrow {}^4F_2$). The sample used was a one-centimeter cube cut from a boule of standard pink ruby . . ., with a concentration of approximately 0.05 weight percent Cr_2O_3 to Al_2O_3. Two components of radiation re-emitted from the crystal were observed in a direction perpendicular to the exciting beam: that due to re-emission of the incoming radiation (spontaneous decay from $^4F_2$) and fluorescence (spontaneous decay from ^2E). The intensity of the first component is proportional to $h\nu_{31}N_3A_{31}$, where [N_3 is the population in level 3 and] A_{31} is the A coefficient for $^4F_2 \rightarrow {}^4A_2$ and is calculated from measurements of absorption coefficient and linewidth for this transition ($A_{31} \sim 3 \times 10^5$/sec). Similarly the fluorescent intensity is proportional to $h\nu_{21}N_2A_{21}$. By a measurement of the ratio of these two components and the use of an auxiliary condition applicable to steady-state conditions $N_2S_{21} = N_3S_{32}$ and also the use of the approximation $S_{21} = A_{21}$, we find $S_{32} \sim 2 \times 10^7$/sec.

"A measurement of fluorescent quantum efficiency, i.e., the number of fluorescent quanta emitted compared to the number absorbed by the crystal from the exciting beam, yielded a value near unity. This result reconfirms the evidence that the life of level 2 is near radiative and also implies that $S_{32} \gg S_{31}$. . . .

"Calculations utilizing the previous results indicated that population changes in the ground state of ruby due to optical excitation would be easily observed. This conclusion was verified in the following experiments. A ruby crystal was mounted between parallel silvered plates to form a microwave cavity resonant at the ground-state zero-field splitting (11.3 kMc/sec). . . . The reflection coefficient of the cavity was monitored on an oscilloscope while a short pulse (200 μsec) of light from a flash tube irradiated the crystal. The magnitude of the microwave magnetic absorption was observed to decrease abruptly and then return to equilibrium with a time constant of about 5 msec. . . . We attribute this effect to temporary depletion of this ground state population with subsequent decay back from the fluorescent level."

The next step was to produce an actual population inversion and stimulated emission. Success in this respect was announced two months after the foregoing developments in *Nature*, reported a bit more extensively the following month in *British Communications and Electronics*, and finally discussed in detail the next year in *The*

Physical Review, in a joint article with R. H. Hoskins, I. J. D'Haenens, C. K. Asawa, and V. Evtuhov.

(As a preliminary to the description, it should be noted that the level designated ^2E in Fig. 13-3 is actually a doublet. The transition from the upper member gives the 6929-Å radiation, which is designated the R_1 line; that from the lower member gives 6943 Å, designated R_2. The energy separation is some 8 times that of the ground state utilized in Maiman's first experiment.)

The problem of a pumping source, while not as serious as Schawlow and Townes had anticipated (they thought in terms of a need for a sharply defined line, whereas the width of the $^4F_2$ band in ruby allowed a range of frequencies), was still a significant one. The authors estimated the intensity required: "If the crystal is illuminated uniformly with isotropic radiation, . . . we find that . . . > 555 watts/cm^2 is required to produce stimulated emission in this crystal. . . .

"Due to the need for high source intensities to produce stimulated emission in ruby and because of associated heat dissipation problems, these experiments were performed using a pulsed light source. For the case in which the exciting light pulses are short compared to the fluorescent lifetime, the requirement on the flash tube is that the energy per unit area is . . . $\cong 1.67$ joules/cm^2.

"The source which was used was a . . . xenon-filled quartz flash tube. . . . [T]he spectral efficiency of the lamp is . . . about 0.064. The radiating area of this source is approximately 25 cm^2 so that an electrical input energy to the lamp of 650 joules would be required to produce stimulated emission in ruby on the basis of the previous considerations.

"The material samples were ruby cylinders about $\frac{3}{8}$ in. in diameter and $\frac{3}{4}$ in. long with the ends flat and parallel to within $\lambda/3$ at 6943 A. The rubies were supported inside the helix of the flash tube, which in turn was enclosed in a polished aluminum cylinder (see Fig. [13-4]); provision was made for forced air cooling. The ruby cylinders were coated with evaporated silver at each end; one end was opaque and the other was either semitransparent or opaque with a small hole in the center.

"A block diagram of the experiment is shown in Fig. [13-5]. The energy to the flashtube was obtained by discharging a 1350-μf capacitor bank and the input energy was varied by changing the charging potential. The R_1 output radiation was monitored with a . . . photomultiplier tube which was calibrated at 6943 A by comparison with [a] . . . thermopile to radiation at this wavelength in a band 200 A wide. The thermopile was calibrated with an NBS standard lamp. The attenuation of the radiation necessary to insure

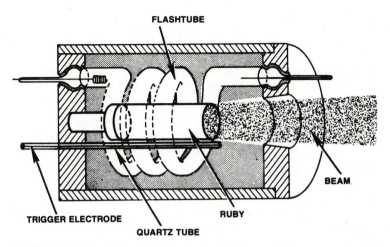

Fig. 13-4. "Apparatus for pulsed excitation of ruby. (Actual size approximately 2 × 1 in. o.d.)" [*Phys. Rev.* **123** (1961), p. 1154, Fig. 7.]

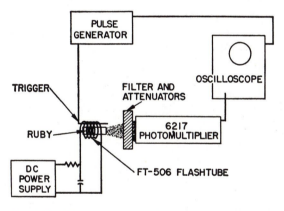

Fig. 13-5. "Block diagram of experimental setup for the observation of stimulated emission in ruby." [*Phys. Rev.* **123** (1961), p. 1154, Fig. 8.]

linear response of the photomultiplier was obtained by the use of calibrated neutral-density gelatin filters. Peak output power and details of the output pulse were obtained from the phototube output across a 1000-ohm resistor on an oscilloscope; the total instrumental response time was about 0.1 μsec. The total energy in the output pulse was obtained by integrating the phototube current with a 0.1-μf mica capacitor.

"It was found with high-intensity excitation that the nature of the output radiation from the various ruby samples which were tried could be divided into two categories:

"(A) Crystals which exhibited R_1 line narrowing of only 4 or 5 times, a faster but smooth time decay of the output (compared to the fluorescence), an output beam angle of about 1 rad, and no clearcut evidence of a threshold excitation. This type of behavior was reported and discussed by Maiman [in the papers in *Nature* and *British Communications and Electronics*].

"(B) Crystals which exhibited a pronounced line narrowing of nearly four orders of magnitude, an oscillatory behavior of the output pulse, and a beam angle of about 10^{-2} rad; these crystals were particularly characterized by a very clear-cut threshold input energy where the pronounced line and beam narrowing occurred."

Maiman and his coworkers expressed the opinion that the instances of small line narrowing and large beam divergence were caused by inhomogeneities and strains in the crystal, which would scatter some of the radiation into undesired modes. The full potentialities of the laser could therefore be realized only with the more nearly perfect crystals. Nevertheless, as indicated above, the achievement of laser action was first achieved with the poorer samples, and quotations from these preliminary reports are of interest.

The note in *Nature* reported only the line narrowing: ". . . [A] ruby crystal of 1-cm dimensions coated on two parallel faces with silver was irradiated by a high-power flash lamp; the emission spectrum obtained under these conditions is shown in Fig. [13-6, part (b)]. These results can be explained on the basis that negative temperatures[12] were produced and regenerative amplification ensued."

The paper in *British Communications and Electronics* also discussed the shortened decay time.

"A series of oscilloscope pictures of R-line decay were taken at progressively increasing pulse intensity. At low power, the decay was approximately a simple exponential with a time constant of 3.8 m.sec. At high pulse power, the decay deviated considerably from the simple exponential with a very fast initial decay constant. With the maximum source intensity available, it was possible to reduce the initial decay constant to 0.6 m.sec.

"We infer from these results that negative temperatures have been obtained and that the lifetime of ^2E was decreased well below the spontaneous value because of stimulated emission."

Very soon thereafter, lasers with several other types of active materials were produced—including, in particular, the gas laser of Javan and coworkers, important because it was the first type of laser that operated continuously rather than in pulses. From these beginnings, the laser has not only developed into a valuable research tool, providing electromagnetic fields of an intensity and a degree of coherence otherwise unattainable at optical frequencies; in addition,

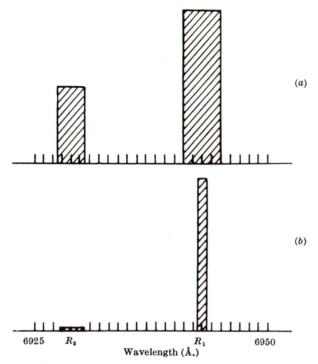

Fig. 13-6. "Emission spectrum of ruby: *a*, low-power excitation; *b*, high-power excitation." [*Nature* **187** (1960), p. 494, Fig. 2.]

it has found application in a number of technologies. The technique as applied to microwaves, on the other hand, has had a more limited range of applications, largely because—as hinted at in the remarks of Gordon, Zeiger, and Townes quoted on p. 219—the very properties that characterize it are such as to make it useful only in specialized situations. But it has also proved highly useful in research; it has growing application in long-distance telephone and television links via satellite; and it is used in essentially all deep-space communications.

Notes

1. *Raman scattering* refers to a process in which a molecule changes its state of vibrational or rotational excitation in the process of interacting with an incident photon; the outgoing scattered radiation is thus shifted in energy relative to the incident.

2. The other half of the prize was split between Nikolai G. Basov and Aleksandr M. Prokhorov of the P. N. Lebedev Institute of the Academy of

Science, Moscow, who independently developed many of the concepts involved in the maser and the laser.

3. This is a common phenomenon in certain classes of molecules and is known as *inversion doubling*.

4. See the discussion of the analogous magnetic situation in Chap. 8 of *Crucial Experiments*.

5. A few obvious typographical errors in the original papers have been corrected in the following excerpts.

6. This is noise arising from the fact that the electron current is not truly continuous but consists of a pulse from each of the electrons as it leaves or reaches an electrode.

7. This notation has to do with the distribution of electric and magnetic fields in the cavity and is of no particular importance in understanding the experiment.

8. Cf. note 10 of Chap. 7. The cutoff condition for a circular wave guide is somewhat different in detail from that for a rectangular wave guide, but the basic concept is similar.

9. The noise figure of an amplifier is defined as follows: There is thermal noise power of $kT\Delta\nu$, where $\Delta\nu$ is the bandwidth of the amplifier, at the input. If the gain of the amplifier is g, there will be a minimum noise power of $gkT\Delta\nu$ at the output. The actual noise power output is always larger than this by some factor F, because of noise produced within the amplifier; F is the noise figure.

10. Spontaneous emission is inherently more troublesome for optical radiation than for microwaves: Other factors being equal, the probability for spontaneous radiation increases as the cube of the frequency, while that for stimulated emission is independent of frequency (this assumes, in both cases, electric dipole radiation).

11. They felt this way despite Schawlow's prejudice expressed later in these words: "Being at Bell Laboratories, I had been pretty thoroughly indoctrinated to believe that anything that you can do in a gas can be done in a solid and can be done better in a solid."

12. This is another way of saying *population inversion*. It refers to the fact that if the distribution function $\exp(-E_i/kT)$ discussed on p. 213 is to give a larger value for a state with higher energy than for a state with lower energy, the value of T must be negative.

Bibliography

The first announcement of the maser is J. P. Gordon, H. J. Zeiger, and C. H. Townes, *The Physical Review* 95, 282 (1954); the detailed papers are J. P. Gordon, *The Physical Review* 99, 1253 (1955), and J. P. Gordon, H. J. Zeiger, and C. H. Townes, *The Physical Review* 99, 1264 (1955).

The considerations on extension of the maser principle to infrared and optical frequencies are given in A. L. Schawlow and C. H. Townes, *The Physical Review* 112, 1940 (1958).

The papers on the laser are T. H. Maiman, *Physical Review Letters* 4, 564 (1960); *Nature* 187, 493 (1960); *British Communications and Electronics* 7, 674 (1960); and T. H. Maiman, R. H. Hoskins, I. J. D'Haenens, C. K. Asawa, and V. Evtuhov, *The Physical Review* 123, 1151 (1961).

Townes's Nobel lecture appears in C. H. Townes, *Science* 149, 831 (1965). Townes also gives some interesting insights into the technological and sociological background in C. H. Townes, *Science* 159, 699 (1968).

Schawlow gives a detailed personal view of the history of the interval from maser to laser in a chapter of *Impact of Basic Research on Technology*, edited by B. Kursunoğlu and A. Perlmutter (Plenum Press, New York, 1973), p. 113.

See also *Lasers and Light* (W. H. Freeman and Company, San Francisco, 1969), chaps. 19-31.

Chapter 14

"Tunneling" and Superconductivity

The discoveries of the zero resistivity and of the perfect diamagnetism of certain metals at low temperatures have been described in Chapter 4. It was pointed out there that the work of Meissner and Ochsenfeld made possible a theoretical description of superconductivity on a megascopic level and from a phenomenological point of view, but that for a long time following their work there was still no understanding of the microscopic mechanism that gave rise to the phenomena.

The lack was overcome in 1957 by the publication of a theory by John Bardeen, Leon N. Cooper, and J. Robert Schrieffer of the University of Illinois. The theory has become known simply as the BCS theory from their initials, and its success led to Nobel prizes for physics in 1972 for its three authors. A sketch of the theory together with some background history is given in Appendix A. The key feature is the existence of a correlation between pairs of electrons. The ground state of a superconductor at zero temperature is one in which all the electrons are paired off under the influence of the correlation, and the energy required to break up a pair gives rise to a gap in the energy spectrum of the electrons. The gap occurs within the conduction band; its width is of the order of kT_c where T_c is the critical temperature for the transition between the normal and superconducting states, and it varies somewhat with temperature, vanishing at $T = T_c$.

In 1959, Ivar Giaever, an engineer at General Electric Research Laboratories, was studying the phenomenon known as "tunneling." This is the name applied to the penetration of a particle through a region which, according to classical consideration, it would be energetically impossible for the particle to enter.[2] A thin insulating film between two pieces of metal constitutes such a region for the electrons in the metal interiors, and this was the sort of arrangement that Giaever was working with. At the same time, he was taking graduate courses in physics at Rensselaer Polytechnic Institute. When he was introduced to the BCS theory, it occurred to him that tunneling could be used as a probe for studying the gap in the electron energy spectrum, and he inaugurated a program to do so. The work provided a large body of useful information and earned Giaever a share of the Nobel prize for physics in 1973.

Then in 1961, a graduate student at Cambridge University, Brian D. Josephson, pointed out that not only single electrons but also pairs could tunnel through sufficiently thin films, and that this possibility would give rise to three new effects: (1) There should exist supercurrents through the films, that is, it should be possible to have a current through a film with zero voltage across it. The maximum possible value of the zero-voltage current should be very sensitive to the application of a magnetic field in the plane of the film, decreasing rapidly (at first) with increasing field. (2) When there is a dc potential V across the film, there should be in addition to the direct current an alternating current of frequency $\nu = 2eV/\hbar$. (3) Conversely, if the applied voltage has both a dc component and an ac component of frequency ν, then the dc voltage-current relation should be modified; in particular, there should be zero-resistance regions (i.e., regions in which the current can be varied with no change in voltage) of magnitude depending on the magnitude of the ac voltage, occurring at values $V_n = nh\nu/2e$, where n is any integer and e is the charge on an electron. Over the next four years all these effects were detected, and several ramifications explored, in a variety of experiments by a number of workers, Giaever among them; and Josephson also won a share of the 1973 Nobel prize in physics.

This chapter describes Giaever's application of tunneling to the study of superconductivity, the verification of Josephson's predictions, and some of the further developments of the effects.

The significance of tunneling for conduction between metals had been recognized, at least qualitatively, for many years. As Giaever and Karl Megerle put it in an Article in *The Physical Review* in 1961 detailing their work with superconductors, "The concept that particles can penetrate energy barriers is as old as quantum mechanics. . . . It has long been known that an electric current can flow between two metals separated by a thin insulating film because of the quantum-mechanical tunnel effect.[3] Theoretical calculations were first made [in 1933] by [Arnold] Sommerfeld and [Hans] Bethe." Experimental studies had also been done, principally by Ragnar Holm and coworkers at the great Siemens electrical company and also by Holm and Walther Meissner; by a student of Meissner's, Isolde Dietrich; and by Meissner's son Hans. The quantitative value of these works is uncertain, however. The contacts were mostly simple spring- or weight-loaded crossed wires, and the nature of the contact surfaces and the films formed on them was not well known. Giaever, with a colleague, John C. Fisher, introduced the procedure of evaporating a strip of metal onto a substrate, allowing an oxide film to form on the surface, and then evaporating a second metal strip across the first. In this way, they were able to obtain thoroughly reliable results, and to give what some regard as the first convincing evidence that the current through an oxide film is indeed due to tunneling.

As mentioned earlier, it was during the course of this work that Giaever learned of the BCS gap and recognized that it might influence the tunneling current. The paper by Giaever and Megerle already quoted gives an excellent presentation of the principles involved, as well as a review of the techniques used in tunneling studies.[4]

"In [part (a) of Fig. 14-1] we show a simple model of two metals separated by a thin insulating film, the insulating film is pictured as a potential barrier. In [part (b)] is shown the case when one of the two metals is in the superconducting state. Note how the electron density of states has changed, leaving an energy gap centered at the Fermi level as postulated by Bardeen, Cooper, and Schrieffer. . . . In [part (c)] both the metals are pictured in the superconducting state.

". . . [W]e shall discuss qualitatively these three different cases. . . .

"The transmission coefficient of a quantum particle through a potential barrier depends exponentially upon the thickness of the barrier and upon the square root of the height of the barrier. For small voltages applied between the two metals neither the barrier thickness nor the barrier height is altered significantly.[5] The current will then be proportional to the applied voltage, because the number of electrons which can flow increases proportionally to the voltage. The temperature effect will be very small, as the electron distribution is equal on either side of the barrier with metals in the normal state and in addition, kT is much smaller than the barrier height.

"When one of the metals is in the superconducting state the situation is radically different. At absolute zero temperature, no

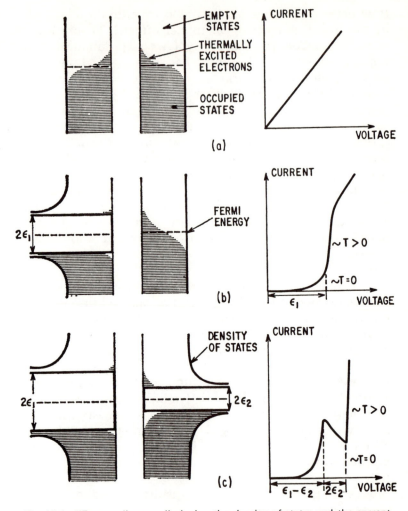

Fig. 14-1. "Energy diagram displaying the density of states and the current-voltage characteristics for the three cases. (a) Both metals in the normal state. (b) One metal in the normal state and one in the superconducting state. (c) Both metals in the superconducting state." The energy level diagrams are for nonzero temperature, so that there are some electrons thermally excited to energies above the Fermi level, which is indicated by the dashed lines. [*Phys. Rev.* **122** (1961), p. 1104, Fig. 4.]

current can flow until the applied voltage corresponds to half the energy gap. Assuming that the current is proportional to the density of states, the current will increase rapidly with voltage at first, and then will asymptotically approach the current-voltage characteristic found when both metals were in the normal state. At a temperature different from zero we will have a small current flow even at the lowest voltages. But since the two sides of the barrier now look different, the current will depend strongly upon temperature.

"When both metals are in the superconducting state, the situation is again different. At absolute zero no current can flow until the applied voltage corresponds to half the sum of the two energy gaps. At a finite [i.e., nonzero] temperature, a current again will flow at the smallest applied voltages. The current will increase with voltage until a voltage equal to approximately half the difference of the two energy gaps is applied. When the voltage is increased further it is possible for only the same number of electrons to tunnel, but since the electrons will face a less favorable (lower) density of states, the current will actually decrease with increasing voltage. Finally, when a voltage equal to half the sum of the two gaps is applied the current will again increase rapidly with voltage and approach asymptotically the current-voltage characteristics obtained when both metals were normal.

"Since we regard the distribution of holes and electrons in both the normal and superconducting states as symmetric about the Fermi level, no rectification effects are expected."

As for apparatus and methods, those described by Giaever and Megerle undoubtedly differed in details from those used by other groups that later[6] became involved in tunneling work, but presumably the essential features must have been common to all.

"The apparatus which is shown in Fig. [14-2] consists basically of a liquid helium Dewar for pumping on the helium,[7] and an outer Dewar containing liquid nitrogen which acts as a radiation shield for the helium. The helium Dewar has a constriction in its diameter to minimize creep losses of the superfluid helium when the temperature is below the λ point.[8]

"Temperatures are measured by means of the helium vapor pressure. . . . The system is capable of attaining a temperature of about 0.9°K. . . . [T]his temperature can be maintained for approximately six hours. . . .

"The electrical circuitry is shown in Fig. [14-3]. To trace out the current-voltage characteristics, a[n] . . . X-Y recorder and matching . . . dc amplifier are used in conjunction with external shunts and multipliers to extend the range of the instruments. . . . [T]he current scale can be decade switched over a full-scale sensitivity 100 ma to 1 μa and the voltage scale from 100 mv to 50 μv."

A variant used by some other groups is worth mentioning: An ac source (usually standard 60-Hz) was used instead of the battery, so as to sweep automatically and repeatedly through the desired current or voltage range. The voltage signal was applied to one of the deflection systems of an oscilloscope, and the current signal to the

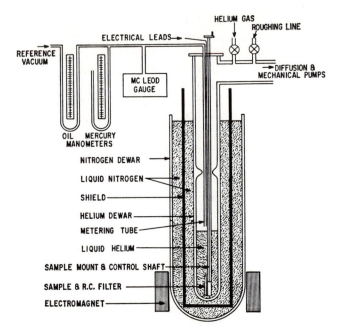

Fig. 14-2. Schematic drawing of the apparatus used by Giaever and coworkers for electron tunneling measurements. "Shield" is a magnetic shield that can be removed to permit studies using a magnetic field. [*Phys. Rev.* **122** (1961), p. 1101, Fig. 1.]

other. Thus the oscilloscope recorded the current-voltage characteristic continuously.

"The emf source can be used as either a high- or low-impedance source, by suitable adjustment of the two variable resistors and the applied battery voltage. The high capacitance of the sample in conjunction with considerable lead inductance gives rise to very troublesome high-frequency oscillations whenever the sample is biased into its negative-resistance region.[9] By placing an adjustable high-pass filter in parallel with the sample, the high-frequency oscillations can be eliminated or at least greatly reduced. The high-pass filter consists of a capacitor large in comparison to the sample capacitance and a variable resistor in series, and is in close proximity to the sample to minimize lead inductance. The variable resistor is . . . mounted on the end of a . . . stainless steel tube. Concentric with this tube is a . . . [smaller] stainless steel tube which engages the adjustment screw of the resistor and passes through an O-ring seal in the Dewar cover plate, to permit external adjustments.

"The electrical connections into the Dewar, consisting of current and voltage leads, are brought out through the cover plate and are sealed in place with Apiezon wax to achieve a tight seal. . . . To

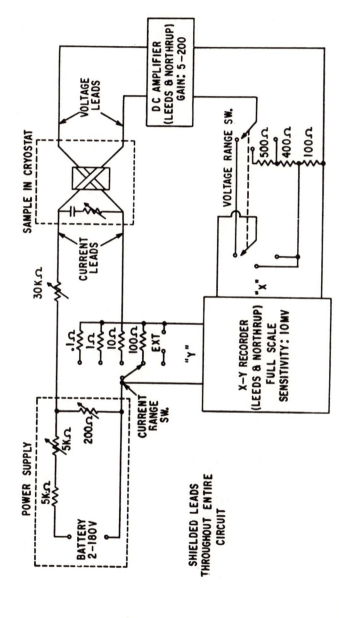

Fig. 14-3. Electrical circuitry as used by Giaever and coworkers for electron tunneling measurements. [*Phys. Rev.* **122** (1961), p. 1102, Fig. 2.]

237

minimize induced noise, the entire electrical circuitry outside the Dewar is shielded. The sample and leads within the Dewar can be shielded by a copper-clad soft iron shield which sits in the liquid nitrogen, surrounding the helium Dewar.

"Since the voltages applied to the sample are very small, induced voltages caused by ever-present fluctuating stray fields remain a difficult problem even after careful shielding. . . . The difficulty is virtually eliminated by including a resistance in series with the current loop, and making this resistance as large as practical. The large resistance in series with the sample resistance acts as a voltage divider for induced noise, so that only a small amount of noise appears across the sample. This . . . cannot be used when investigating the negative-resistance region of the sample. It has, however, been retained for measurements outside the negative resistance region on most samples.

"The sample is mounted directly on the variable resistor which is a part of the high-pass filter. This insures mechanical rigidity and reproducible, accurate positioning for measurements involving magnetic fields. . . .

"The sample consists of two metal films separated by a thin insulating layer. Aluminum/aluminum oxide/metal sandwiches are prepared by vapor-depositing aluminum on microscope glass slides in vacuum, oxidizing the aluminum, and then vapor-depositing a metal over the aluminum oxide. First the microscope slide is cut to size, $\frac{1}{2} \times 3$ in.,[10] so as to fit through the constriction in the helium Dewar. Next, indium is smeared onto the four corners of the glass slide to provide contacts between the evaporated metal strips and external leads. The glass slide with indium contacts is then washed with . . . detergent, rinsed with distilled water and ethanol, and dried with dry nitrogen gas.

"Next, the glass slide is mounted in the evaporator so that it can be positioned behind suitable masks in vacuum. The evaporations are made from tantalum strips . . . which have previously been charged [with the metal to be evaporated] and heated in vacuum so that the charge wets the tantalum strip. The evaporations are made at a starting pressure of 5×10^{-5} mm Hg, or less.

"Preparation of the metal/insulator/metal sandwich proceeds in three distinct steps, as shown in Fig. [14-4], during which the substrate is at room temperature.

"First, a layer of aluminum is evaporated onto the glass slide between two contacts. This strip is 1 mm wide and 1000-3000 Å thick. Next, the aluminum is oxidized either at atmospheric pressure or some reduced pressure. Finally, a layer of Al, Pb, In, or Sn, of

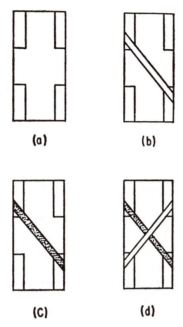

(a)   (b)

(C)   (d)

Fig. 14-4. "Sample preparation. (a) Glass slide with indium contacts. (b) An aluminum strip has been deposited across the contacts. (c) The aluminum strip has been oxidized. (d) A lead film has been deposited across the aluminum film, forming an Al-Al$_2$O$_3$-Pb sandwich." [*Phys. Rev.* **122** (1961), p. 1103, Fig. 3.]

dimensions similar to the aluminum strip, is evaporated over the aluminum oxide layer between the remaining two contacts.

"The thickness of the Al$_2$O$_3$ insulating layer between the metal strips is subject to a number of variables. . . . [Those discussed are the pressure and time of oxidation, water vapor in the oxidizing atmosphere, evaporation rate and evaporation temperature of the overlying metal, oxidation temperature, and rate of evaporation of the aluminum layer.]

"By controlling these parameters to some extent, the resistance of a 1-mm^2 junction can be made to vary between 10^{-2} and 10^7 ohm. . . .

"It is possible to measure indirectly the thickness of the oxide layer by measuring the capacitance of the junction and then calculating the thickness."

In his first report, in *Physical Review Letters* in August 1960, Giaever showed measurements of characteristics like that in part (b) of Fig. 14-1. In addition, he pointed out that the value of the slope of the curve of current versus voltage, at any given value of voltage, was proportional to the number of available electron energy states per unit energy in the superconductor; this was as expected on the basis of what Giaever referred to as a "naive picture" of the theory of tunneling, but few theorists at the time had been confident of its validity. It received further confirmation three months later, when both Giaever and a group from Arthur D. Little, Inc., composed of James Nicol, Sidney Shapiro, and Paul H. Smith, reported in successive Letters in *Physical Review Letters* the measurement of the negative-resistance region as shown in part (c) of Fig. 14-1. Finally, in the Article quoted above, Giaever and Megerle presented a number of measurements of both types, on a variety of superconductors, and at a number of different temperatures and magnetic fields, all of which dovetailed beautifully with the BCS theory.

All these results, however, pertained to the tunneling of free (unpaired) electrons and examined the distribution of single-particle energy states, particularly the gap. The phenomena predicted by Josephson, on the other hand, are consequences strictly of the existence and properties of pairs (see Appendix A) and particularly of the inapplicability of the exclusion principle to pairs. If a barrier between two superconductors has a thickness that is small compared with the decay length of the electron wave function in the barrier,[11] then the pair function from each side can penetrate the barrier to the other side. This implies, first, the tunneling of pairs themselves through the barrier, and thus the possibility of a pair current, which is not inhibited (as the single-particle tunnel current is) by the occupancy of states on the exit side. Next, there is implied a relationship between the phase functions θ on the two sides of the barrier.

A full mathematical treatment is far too complex for presentation here, but the basic concepts are fairly simple. As pointed out in Appendix A, the phase function θ has a single value throughout a superconductor (as long as there is neither a supercurrent nor a magnetic field present), but the phase can be changed everywhere by the same amount without physical consequences. In a system of two superconductors separated by a large insulating barrier, the phase of each can be altered in this manner independently of the other. As the thickness of the insulating barrier is reduced to zero, however, the properties of the system may be reasonably assumed to change continuously into those of a single superconductor. This means that there is an energy of interaction, or coupling, between the two that depends on the difference in the phases and that increases in magnitude, becoming more negative, as the thickness of the barrier decreases. In the limit of weak coupling, the full theory shows that the energy per unit area has the form

$$E_c = -E_0 \cos \delta,$$

where $E_0 > 0$ depends on the structure of the junction[12] and δ is the difference between the values of θ on the two sides of the junction. The variation of θ from one point to another in turn implies (see Appendix A) a supercurrent per unit area of the form

$$j = j_1 \sin \delta, \tag{14-1}$$

where $j_1 = -2eE_0/\hbar$. The value of δ is partly but not directly under the control of the experimenter. For one thing, it is affected by the presence of a voltage across the barrier. With zero voltage, however, it is constant; then the experimenter can vary the current, by means of the circuitry external to the junction, within the limits imposed by Eq. (14-1).

As mentioned earlier, Josephson arrived at his results while he was a graduate student at Cambridge; his interest in the problem had been aroused by a course of lectures by P. W. Anderson, who was spending a year at Cambridge on a sabbatical leave from Bell Telephone Laboratories, and he naturally discussed the problem with Anderson.[13] As Anderson tells it in an Article in *Physics Today*, "After a few more discussions at the Cavendish [Laboratory at Cambridge] ... I returned home to Bell Labs. There I mentioned to [John M.] Rowell my conviction that Josephson was right. Rowell admitted to me that he had, from time to time, seen suggestive things in his tunnel junctions, and a few months later he called me in to look at his experimental results on a new batch of tunnel junctions. He thought he might actually be seeing the Josephson effect." After a number of checks to verify that it was not spurious, Anderson and Rowell published the report of their observation in *Physical Review Letters* in March 1963.

"We have observed an anomalous dc tunneling current at or near zero voltage in very thin tin oxide barriers between superconducting Sn and Pb, which we cannot ascribe to superconducting leakage paths across the barrier, and which behaves in several respects as the Josephson current might be expected to.

"Figure [14-5] shows an *X-Y* recorder plot of the tunneling current vs voltage for one of these structures at $\sim 1.5°$K. The lead and tin films are both approximately 2000 Å thick, and the junction has dimensions 0.025×0.065 cm^2 and a resistance (both metals normal) of 0.4 Ω. Voltage is applied to two arms of the junction from a 1-kΩ potentiometer and the resulting current flow is measured as voltage across a series resistor of 10 Ω. The voltage appearing

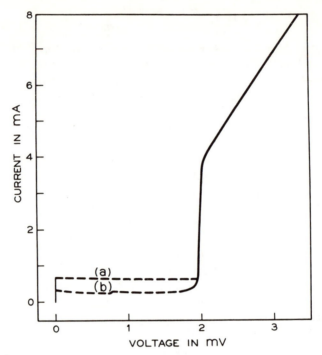

Fig. 14-5. "Current-voltage characteristic for a tin-tin oxide-lead tunnel structure at ~1.5°K, (a) for a [magnetic] field of 6 × 10⁻³ gauss and (b) for a
field 0.4 gauss." [*Phys. Rev. Lett.* **10** (1963), p. 230, Fig. 1.]

across the barrier is taken directly from the other two arms of the
junction. Figure [14-6] shows the plot with current scale expanded
to show the anomalous region near the origin. The current at first
increases up to a value of 0.3 mA with no voltage appearing across
the barrier. At this point the junction becomes unstable and may
fluctuate back and forth between the vertical characteristic and the
expected 'two-superconductor' characteristic.[14] With a small increase
in current, the junction settles stably on the latter. . . .

"One possible explanation that will be suggested is, of course,
that in such thin junctions we have not avoided small superconducting shorts across the barrier. There are, however, four experimental
points suggesting that this is indeed the Josephson effect.

"(1) As pointed out in Josephson's Letter, the effect should be
quite sensitive to magnetic fields. . . . [W]e expect an additional
dependence on . . . the vector potential, which will lead to cancella-
tion of currents in various parts of the barrier if the magnetic flux
flowing between the superconductors reaches one or two quanta.[15]
With an area of ~10⁻⁷ cm², this corresponds to about 1 gauss. We
have found (see Fig. [14-6]) that when the junction was carefully

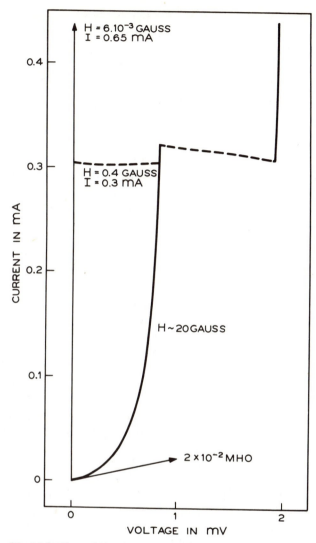

Fig. 14-6. Expanded version of the low-current portion of Fig. 14-5, together with a curve showing the characteristic at a field of ~20 gauss. [*Phys. Rev. Lett.* **10** (1963), p. 230, Fig. 2.]

shielded by a mu-metal can with a measured interior field of 6×10^{-3} gauss, the vertical characteristic reached 0.65 mA, with no shielding (0.4 gauss), 0.30 mA, and when less than 20 gauss was applied the anomalous behavior was not observed. Fine superconducting filaments should show anomalously high, not low, critical fields."

The other checks will not be described in detail; they involved the presence of the effect only when both metals were superconducting, the actual conductivity of the barrier, and the behavior under attempted "burnout" of a supposed filament. Anderson and Rowell conclude, "These . . . arguments seem to us nearly to exclude the conducting leak hypothesis."

Actually, Josephson himself had made a junction and tried to observe the effect, without success. The problem is that unless the coupling energy E_c is larger than the noise energy in the circuit, the correlations will be broken by thermal fluctuations and the Josephson effect will be masked. Thus the quality of Rowell's junctions, and in particular their low resistance,[16] were essential to the success of the experiment. Anderson comments, however, that others had probably seen the effect but failed to recognize what was happening.[17]

The magnetic test that Anderson and Rowell describe had been a bare minimum.[18] It could be expanded into a much more significant verification of the true nature of the effect, based on the fact that the dependence of δ on the magnetic field (see Appendix A, page 297) turns out to be quite sensitive. If there is a uniform magnetic field in the plane of the barrier, δ will increase linearly in the direction of the field. There can then be interference between the currents from different parts of the junction, just as there is interference, at a screen, of light rays from different parts of a slit; and a change of the magnetic field in the barrier is completely analogous to a change of the angle of observation of the slit in the optical case. Rowell, therefore, studied this effect in more detail and reported the results together with other studies in *Physical Review Letters* in September 1963.

"It has been observed by Josephson that the effect of an external magnetic field . . . would be to reduce the direct current to a minimum whenever the junction contained integral numbers of flux units ($hc/2e = 2.1 \times 10^{-7}$ gauss cm^2). We have observed this effect in junctions of various dimensions, but it is most striking when the film along the field direction is as narrow as possible. Figure [14-7] shows the variation of observed current as a function of magnetic field for a Pb-I-Pb junction[19] made of films 0.040 and 0.24 mm wide, the 0.40-mm wide film being roughly along the field direction. When a field of 6.5 gauss is applied, the Josephson current is reduced by a factor >600 and cannot be measured with the existing experimental sensitivity. At 13.0 and 19.5 gauss, the current again goes through minima with successively decreasing maxima between. The area of the junction containing flux is the width $W \times 2\lambda = 3.1 \times 10^{-8}$ cm^2 [here λ is the penetration depth; see Appendix A, page 297]. . . . Thus a field of 6.5 gauss corresponds to a flux of 2.0×10^{-7} gauss cm^2 in the junction, which is indeed the flux unit. Considering screening and demagnetizing effects of the films, it is surprising that such a good value is obtained."

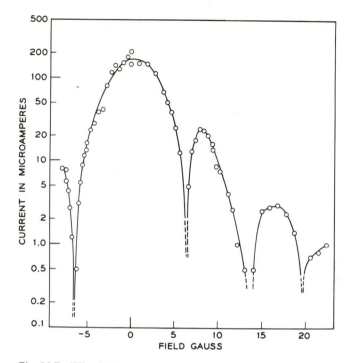

Fig. 14-7. "The field dependence of the Josephson current in a Pb-I-Pb junction at 1.3°K." The similarity to a single-slit optical diffraction pattern is evident. [*Phys. Rev. Lett.* **11** (1963), p. 202, Fig. 3.]

Meanwhile, the first report on the observation of an ac Josephson effect had appeared. There are actually a variety of such effects, depending on the nature of the applied electric (and sometimes also magnetic) field, and only a small sample can be discussed here; the choice is primarily on the basis of simplicity of interpretation. It is first necessary to examine how an applied voltage affects the Josephson current.

The key point is that the time rate of change of the phase θ of the pair function contains a term dependent on the electrostatic potential (see Appendix A, page 297). In particular, a constant potential difference V across a barrier in a tunnel junction makes the phase difference δ in Eq. (14-1) take the form

$$\delta = \delta_0 + (2eV/\hbar)t,$$

and the Josephson current becomes an alternating current of frequency $2eV/\hbar$. The factor $2e/\hbar$ has the value 0.484 GHz/μV, so that for reasonable values of V (of the order of microvolts) the frequency is in the gigahertz range and would not show up in the dc experiments of Rowell. If a suitable coupling between the junction and the surroundings can be effected, however, this direct ac effect can be detected by microwave methods. As will be described later, this has indeed been accomplished; but other versions came first.

If there is both a dc and an ac component of the potential, $V = V_0 + V_1\cos\omega t$, the situation becomes somewhat more complicated. Now j has the form

$$j = j_1\sin[\delta_0 + (2eV_0/\hbar)t + (2eV_1/\hbar\omega)\sin\omega t]: \qquad (14\text{-}2)$$

The alternating current due to the dc voltage is frequency modulated at the frequency of the applied ac voltage. It can therefore be expressed as a sum of terms, each term oscillating at a frequency that is a sum or difference of the Josephson frequency $2eV_0/\hbar$ and an integral multiple of the applied frequency ω, or $\omega' = 2eV_0/\hbar \pm n\omega$.[20] If ω is a submultiple of $2eV_0/\hbar$, then, the oscillatory part of one term vanishes, leaving a term proportional to $\sin\delta_0$, which will constitute a dc component. Two points must be noted: (1) The value of δ_0 is not fixed by external conditions other than the current, so the current can vary at constant voltage (though it is still limited by the fact that δ_0 can range only between $\frac{1}{2}\pi$ and $-\frac{1}{2}\pi$); (2) this effect may be superimposed on the single-particle tunneling current of the kind originally studied by Giaever, and so the resulting "steps" need not be symmetrically located around zero current.

It was this latter effect that was observed by Sidney Shapiro and reported in *Physical Review Letters* in July 1963.

"In the course of experiments on the effects of microwave fields on superconducting tunneling, we have had occasion over the past few months to fabricate many tunneling crossings of low resistance (5−20 Ω with a crossing area of 1.5×10^{-4} cm²). . . . Our experiments have brought to light several new effects which we summarize below.

"The samples were $Al/Al_2O_3/Sn$. Two five-mil-wide Al lines, evaporated onto cleaned glass substrates, were oxidized in a glow discharge. . . . A five-mil-wide cross-strip of Sn was then evaporated forming two samples on each substrate.

"The tunneling current versus voltage characteristics were displayed on an *X-Y* oscilloscope. A low-impedance source was used to drive the loop containing the sample and the current-measuring resistor. . . . The source was either dc, ac, or both in combination. . . .

"The effect of microwave power in modifying the *I-V* curve, especially the zero-voltage currents, was . . . studied. The samples were mounted in a microwave cavity resonant, at low temperatures, at about 9300 Mc/sec and 24 850 Mc/sec. All the following observations were independent of sweep frequency and were also seen when dc was passed through the sample.

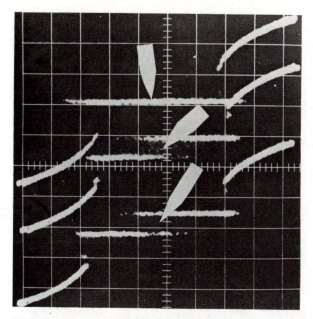

Fig. 14-8. "Initial effect of microwave power" on the *I-V* characteristic of a tunneling junction. Current is plotted horizontally, voltage vertically. "Pointers mark origin." [*Phys. Rev. Lett.* **11** (1963), p. 81, Fig. 2.]

"(1) Figure [14-8] shows for a typical sample the initial effect of 9300-Mc/sec microwave power on the *I-V* characteristic.[21] Without applied power, the trace (top) is similar to that . . . [showing the dc effect]. With a few tens of microwatts applied, however, the zero-voltage currents become noisy [middle trace] and gradually vanish resulting in regions of zero slope (or almost zero slope) in which the current rises at (or almost at) fixed voltage across the sample [bottom trace]. The voltage at which the zero-slope regions occur is equal to $\pm h\nu/2e$, where ν is the microwave frequency. . . . As more power is applied (not shown), further zero-slope regions appear at still higher voltage. . . . The interval in voltage from one zero-slope region to the next is not always $h\nu/2e$; sometimes a step is missing so that the voltage interval is $h\nu/e$.

"(2) Figure [14-8] demonstrates still another startling feature of the effect of microwaves on the Josephson currents. In the bottom trace the origin has vanished as a stable state, and the system is able to remain biased at, e.g., $+h\nu/2e$, not only when positive current is flowing but even when the current is zero and, more astonishingly, *even when the current is reversed and made negative.* . . .

"(3) Similar effects occur at 24 000 Mc/sec.[22] Fig. [14-9] shows the numerous steps at 9300 Mc/sec (A) and at 24 850 Mc/sec (B) which are present at intermediate power levels.

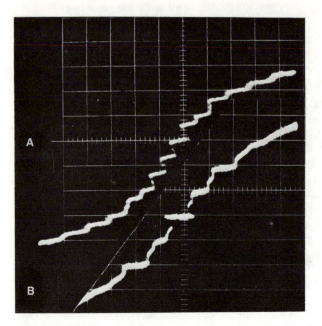

Fig. 14-9. Zero-slope regions in the current-voltage characteristics of tunnel junctions, produced by applied microwave power at 9300 Mc/sec *(A)* and 24 850 Mc/sec *(B)*. "For *A*, *hν/e* = 38.5 μV, and for *B*, 103 μV. For *A*, vertical scale is 58.8 μV/cm . . .; for *B*, vertical scale is 50 μV/cm." [*Phys. Rev. Lett.* **11** (1963), p. 82, Fig. 3.]

"Josephson . . . predicted the occurrence of zero-slope regions separated by *hν/2e* in the *I-V* characteristic in the presence of the rf field. . . . Our experiments have confirmed this prediction and represent indirect proof of the reality of Josephson's ac supercurrent."

Other indirect means of observing the ac effect were devised. For example, the junction layer itself can act as a microwave cavity; if it is tuned to the Josephson frequency by varying the dc voltage and an applied magnetic field, the resulting standing wave produces observable effects on the dc current. Such an effect was noticed in 1963 by Milan D. Fiske of General Electric Research Laboratories; Fiske thought it "possible that the steps are associated with modes in the alternating currents which are developed across the junction at nonzero dc voltages as required by Josephson," but confessed that the matter was "not yet understood in detail." The effect was definitely established in 1965 by R. E. Eck, D. J. Scalapino, and B. N. Taylor of the University of Pennsylvania and by D. D. Coon and Fiske at G.E. Direct observation of the ac effect, however, was complicated by the difficulty of coupling the very high-frequency radiation out of the junction layer. In May 1965, Giaever reported in *Physical Review Letters* a degree of success in this problem.

". . . I have succeeded in detecting the electromagnetic fields [radiated by the junction], using a highly unconventional spectrum analyzer. [Aly] Dayem and [R. J.] Martin [of Bell Telephone Laboratories] showed that when a conventional tunnel junction of two equal superconductors is subjected to a microwave field, current steps occur at voltages given by $(1/e)\ (2\Delta \pm nh\nu)$, where ν is the frequency of the microwave field and 2Δ is the energy gap of the superconductors.[23] Thus a superconducting tunnel junction is indeed a crude spectrum analyzer.

"Now consider the following experimental arrangement shown schematically in Fig. [14-10]. First a Sn film (marked 1 in the figure) is evaporated onto a microscope glass slide. This film is oxidized in the laboratory air overnight to form a thick oxide. Second, another Sn film (2) is evaporated over the first one, forming a T-like structure. This film is oxidized from 5 to 30 min to form a thin oxide layer, and finally a third Sn film (3) is evaporated on top of the other two films. The films (1) and (2) are separated by an oxide layer thick enough to quench out most, if not all, Josephson effects. These two films comprise the detector. Films (2) and (3) are separated by a thin oxide and exhibit the Josephson effects. These two films act as the generators of the microwaves. There also is a 'sneak path' between films (3) and (1); however, the area of this junction can be made small enough so that it does not interfere appreciably with the

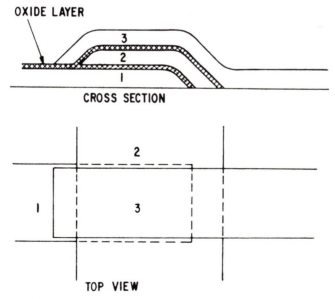

Fig. 14-10. Schematic drawing of the arrangement used by Giaever to detect the ac Josephson effect. All the films are tin. [*Phys. Rev. Lett.* **14** (1965), p. 905, Fig. 1.]

dc measurements. This overlap of film (3) on film (1) is necessary to obtain a tight coupling between the two cavities, represented by the oxide layers between films (1) and (2) and films (2) and (3).

"By applying a voltage V_{23} across the films (2) and (3), a microwave field of frequency $h\nu = 2eV_{23}$ is produced. Then by measuring the current-voltage characteristic of the junction between films (2) and (1), current steps are obtained at voltages $V_{12} = (1/e)$ $\times (2\Delta \pm n h\nu) = (1/e) (2\Delta \pm n2eV_{23})$. . . . One of the better samples is shown in Fig. [14-11], where steps can be seen at least to $n = 3$. . . .

". . . [T]his experiment helps confirm the Josephson ac effect. An appreciable amount of ac power is produced when a junction is put in this mode; to date the order of 10^{-7} W have been extracted from the generator without affecting it to any appreciable degree."

A still more direct demonstration was provided in March 1965 in a Letter to the Editor of the Russian *Journal of Experimental and Theoretical Physics* by I. K. Yanson, V. M. Svistunov, and I. M. Dmitrenko of the Physico-technical Institute for Low Temperatures of the Ukrainian Academy of Sciences, reporting the detection by standard techniques of microwave radiation at the Josephson frequency emitted by a junction. Eck, Scalapino, and Taylor, together with D. N. Langenberg, had been attempting the same method virtu-

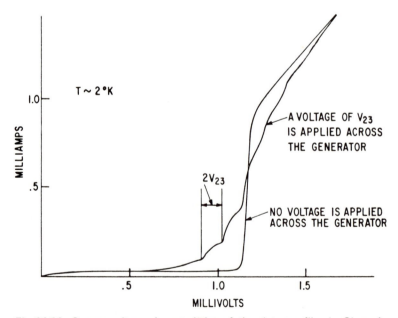

Fig. 14-11. Current-voltage characteristics of the detector film in Giaever's experiment, with and without a voltage applied across the generator film. [*Phys. Rev. Lett.* **14** (1965), p. 905, Fig. 2.]

ally ever since their success with the indirect process. The success of Yanson, Svistunov, and Dmitrenko convinced them that their difficulty was an inadequately sensitive receiver; they therefore built a new one, and reported success in *Physical Review Letters* in August 1965. Their paper gives interesting experimental details.

"In Fig. [14-12], a cutaway view of the wave-guide–junction–magnetic-field configuration is shown. The axis of the wave guide defines the z axis. The externally applied dc magnetic field H_0 was oriented in the plane of the junction perpendicular to this axis producing a spatial modulation of the Josephson ac supercurrent along the length L of the junction (Fig. [14-12, inset]). This supercurrent excites . . . waves which propagate in the insulating layer parallel to the z axis with a phase velocity $\bar{c}$. Thus, a junction of length L has characteristic frequencies given by $\omega_n = n\bar{c}/L$ ($n = 1, 2, 3, \ldots$). The excitation of these modes gives rise to steps in the I-V characteristics[24] at bias voltages $V_n = \hbar\omega_n/2e$. . . . For the junctions used in these experiments, $L = 0.16$ cm and $\bar{c} = 0.05c$ (c is the velocity of light), so that the $n = 2$ mode occurred at 9.2 Gc/sec. The insert in Fig. [14-12] shows the spatial variation of the electric field for this mode

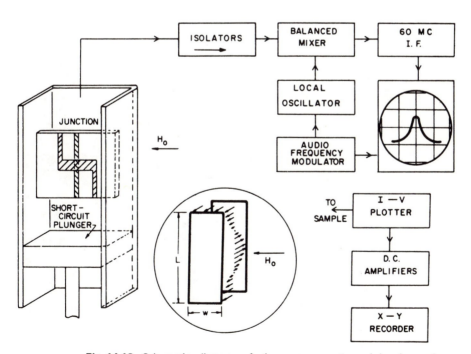

Fig. 14-12. Schematic diagram of the arrangement used by Langenberg, Scalapino, Taylor, and Eck for the direct detection of the ac Josephson effect. [*Phys. Rev. Lett.* **15** (1965), p. 294, Fig. 1.]

which was chosen so that the electric fields at the ends of the junction were in phase and would contribute coherently.

"A simple estimate of the magnitude of the power radiated by the junction . . . gives $P_{rad} \approx 5 \times 10^{-12}$ W.

"In order to investigate this radiation, samples were mounted in a wave guide containing a short-circuit plunger (Fig. [14-12]) which could be adjusted from outside the cryostat. The radiation emitted by the sample passed . . . to an X-band superheterodyne receiver . . . which used a 60-Mc/sec i.f. amplifier with a bandwidth of 4 Mc/sec. The larger signals were displayed directly on an oscilloscope by using a 100 cps sinusoidal voltage to modulate the local oscillator (LO) frequency and also to drive the X axis of the oscilloscope. The video output signal of the i.f. amplifier was applied to the Y axis of the oscilloscope.

"The initial observations of radiated power were made with a junction . . . biased on the $n = 2$ mode. Fig. [14-13] shows a signal observed for this case corresponding to a power level of 10^{-12} W. . . .

"The junction voltage at which signals appeared were found to obey . . . the equation

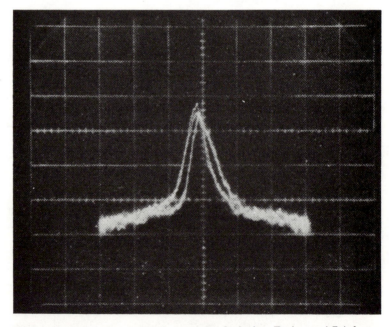

Fig. 14-13. Signal observed by Langenberg, Scalapino, Taylor, and Eck from a Josephson junction radiating at 9.2 GHz. The total width of the trace is about 25 Mc/sec. "The peak results from sweeping the difference frequency between the LO and the radiation from the junction through the 4-Mc/sec pass band of the i.f. amplifier." [*Phys. Rev. Lett.* **15** (1965), p. 295, Fig. 3.]

$$V = \frac{h}{2e}\left(\nu_{\mathrm{LO}} + \nu_{\mathrm{i.f.}}\right), \tag{[14-]3}$$

where ν_{LO} is the LO frequency and $\nu_{\mathrm{i.f.}}$ is 60 Mc/sec. . . . Whenever the LO frequency was changed the bias voltage had to be reset according to Eq. ([14-]3), and the two signals were always separated by a voltage corresponding to a frequency difference of about 120 Mc/sec. These facts leave little doubt that the junction was actually radiating."

By this time, there could be no question as to the validity of Josephson's ideas. There have been a number of applications and extensions of the basic arrangement, two of which are of enough intrinsic interest to warrant brief description.

At California Institute of Technology, James Mercereau had been interested for some time in long-range phase coherence and related effects; he and a graduate student, Lorin Vant-Hull, had made an unsuccessful attempt to detect flux quantization.[25] In 1962, Mercereau learned of Josephson's work and recognized that the effect was indeed a quantum phase detector. By this time he had moved to Ford Research Laboratories. There he quickly acquired three collaborators: Robert Jaklevic, John Lambe, and A. H. Silver. They concentrated their attention on effects arising from the use of two junctions in parallel. Their first result, published in *Physical Review Letters* in February 1964, was a straightforward analog of the optical double-slit diffraction pattern, the additional periodicity corresponding to the presence of two slits being determined by the flux enclosed between the junctions. The second result, published a month later in the same journal, was more interesting: the measurable effect of a vector potential in circumstances such that no magnetic field was present in the junctions themselves. Both experiments were discussed in more detail in an Article in *The Physical Review* in November 1965, from which the following extracts are taken.

The paper begins by deriving the expression for the total current [not current density as in Eq. (14-1)] in a single junction:

$$\text{"}I = I_0 \frac{\sin(\pi\Phi_j/\Phi_0)}{\pi\Phi_j/\Phi_0} \sin\delta(0),$$

where Φ_j . . . is the flux enclosed by the effective cross-sectional area of the junction, $\Phi_0 = h/2e \approx 2.1 \times 10^{-7}$ G cm^2 and $I_0 = j_0\sigma$. [j_0 is the same as the coefficient J_1 in Eq. (14-1), and σ is the area of the junction.] The phase difference $\delta(0)$ adjusts to the experimental conditions." The paper then considers the case of two junctions in parallel. "The total current will now be

$$I = I_{10}' \sin \delta_1 + I_{20}' \sin \delta_2,$$

where δ_1 and δ_2 are the phase differences across the junctions . . . and the diffraction effects are contained in I_{10}' and I_{20}'. . . .

"If (1) and (2) are connected by superconducting links, δ_1 and δ_2 are not random but are related . . . as long as phase coherence persists. . . . [T]he maximum supercurrent flow through the junction pair ('interferometer') is

$$I_{\max} = 2I_0 \frac{\sin(\pi\Phi_j/\Phi_0)}{\pi\Phi_j/\Phi_0} \cos(\pi\Phi_T/\Phi_0).$$

"The total flux Φ_T enclosed by the circuit has as its sources the flux due to external field Φ_a, and also an inductive contribution due to the currents themselves. . . . [For many cases this latter effect] will be very small and it is permissible to set the enclosed flux Φ_T equal to Φ_a."

Three types of experiments were carried out with such devices, only one of which, the second, will be discussed here.

"The second experiment . . . will be the use of the interferometer as a flux meter in the absence of an applied field. In 1949, [W. E.] Ehrenberg and [R. E.] Siday pointed out that interference patterns observed in an electron microscope should depend on the magnetic flux enclosed by the accessible paths, even when there is no magnetic field available to the electrons. . . . This was later discussed extensively by [Y.] Aharonov and [D.] Bohm who . . . pointed out that the effect is nonclassical in origin and arises from . . . [mathematical conditions] imposed on the wave function. Their conclusions have opened discussions concerning the physical significance of the vector potential **A**. . . .

"It has been pointed out that an analogous situation occurs for multiply connected superconductors and that macroscopic quantization is another example whereby the behavior of a charged particle depends on flux not directly accessible to the particle. In this connection if the ideas regarding the coherence of the superconducting ring were true, it should be possible to perform the interference experiments previously done with electron beams with the double-junction superconducting 'interferometer.' In place of the uniform applied field we substitute a long thin solenoid to confine all the flux inside the coil and away from the superconducting circuit. In this case, since the field is confined to the solenoid, there is no flux change in the junctions themselves. . . . The effect expected by this technique will be nondiffraction modulated interference produced directly by the static vector potential. . . .

"The experimental technique for fabrication of the tin–tin oxide–tin tunnel junction differs in some details from previous methods. The base tin layer was deposited in a vacuum of 10^{-6} Torr on [cooled,] cleaned glass of fused silica substrates with standard commercial grade polished surfaces. . . . After deposition of the base film, the substrates were warmed to room temperature in vacuum and removed for further processing. The base film was masked with a plastic film of Formvar[26] to delineate the junction areas and build up the thick areas enclosed by the final junction pairs. . . . After drying the Formvar film for several hours the oxide was thermally grown on the exposed junction regions by exposure to dry flowing tank oxygen for 1 hour at 110°C. The second, top film was then deposited in vacuum at room temperature. . . .

"The small solenoids were constructed by closely winding a fine insulated copper wire around a beryllium-copper core with the core providing the return path. . . . One type was of 1-mil wire wound on a 3 mil core giving an over-all diameter, including insulation, of 6 mils. A second size was similarly constructed of $\frac{1}{2}$ mil wire around a 1 mil core, with an over-all diameter of 2.2 mils. . . . The coils were built into a junction pair device by embedding them in the plastic insulation enclosed by the superconducting interferometer. At least 5 μ of plastic insulation separated the coil from the superconducting circuit. The second tin films were then deposited over the plastic in the usual manner." A cross sectional view of such a device is shown in Fig. 14-14.

". . . [F]our successful devices were made which allowed the observation of periodicity when the flux through the solenoid was varied. As a best example, the curves shown in Fig. [14-15] exhibit the modulation due to the coil itself $\Phi(A)$ and on the same vertical scale that due to an applied field $\Phi(B)$. No diffraction effects are

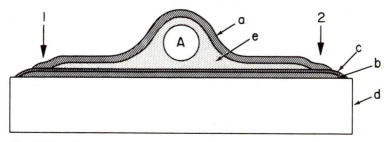

Fig. 14-14. "Cross-section of a Josephson junction pair vacuum-deposited on a quartz substrate (d). A thin oxide layer (c) separates thin (~1000 Å) tin films (a and b). The junctions (1) and (2) are connected in parallel by superconducting thin film links enclosing the solenoid (A) embedded in Formvar (e). Current flow is measured between films a and b." [*Phys. Rev. Lett.* **12** (1964), p. 274, Fig. 1.]

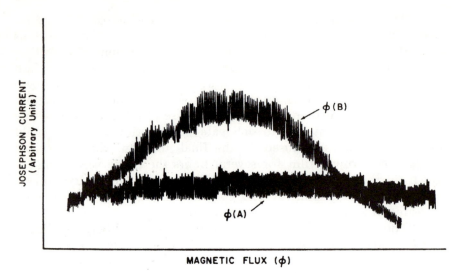

JOSEPHSON CURRENT (Arbitrary Units)

ϕ(B)

ϕ(A)

MAGNETIC FLUX (ϕ)

Fig. 14-15. "Experimental trace of I_{max} versus applied flux for a junction pair showing modulation due to an applied field $\Phi(B)$ and vector potential above [*sic;* presumably a typographical error for "alone"] $\Phi(A)$." [*Phys. Rev.* **140** (1965), p. A1636, Fig. 9.]

apparent with the coil modulation, as expected. . . . For the coil, the periodicity in flux as measured by the current exciting the coil was 16.9 μA per period, while the periodicity in field was 1.2 mG. From the coil calibration, this gives a value of the flux unit of 2.1 $\pm$ 0.1 $\times$ 10^{-7} G cm^2. . . . Attempts were made to determine how much, if any, flux leakage occurred outside the coils. . . . [The result was] an upper limit of 0.01% of the field needed to produce the shift by means of a leakage field."

The result gave a clear demonstration of the macroscopic character of superconducting phase coherence, as did the other two types of experiment. In addition, it confirmed the special importance of the vector potential in quantum phenomena.

At the University of Pennsylvania, after their successful detection of the Josephson radiation, Taylor and Langenberg had set about looking for ways that it could be utilized. It occurred to them that the effect would be an ideal method for measuring the combination of fundamental constants e/h: All that was needed was to measure a voltage and a microwave frequency. Taylor was enthusiastic enough that he arranged to postpone for a year a shift to RCA Laboratories; and he and Langenberg enlisted the assistance of a graduate student, William H. Parker. A preliminary experiment was carried out using an unsaturated standard cell and a borrowed microvolt potentiometer and photocell galvanometer. Three methods were used: direct detection of the radiation from an appropriately biased junction, measurement of the dc voltage induced in an unbiased junction by microwave irradiation, and measurement of the "step" voltages produced in the *I-V* characteristic by microwave irradiation.

The results of this experiment, published in *The Physical Review* in October 1966, were promising enough that a full-scale effort was launched to obtain a truly high-precision result. The third of the three methods was chosen, as it had proved most accurate. A bank of saturated standard cells was purchased and taken to the National Bureau of Standards for calibration; Taylor, in a letter, has recalled the automobile trips between the manufacturer in Rhode Island and the Bureau in Washington, D.C., "with the standard cell enclosure[27] operating off the cigarette lighter and Bill [Parker] and I fretting over the bumps in the road and the weather forecasts—if the temperature approached 90°F, the cells were in trouble." The other problem was that of measuring a millivolt to a precision of a nanovolt with an accuracy of one part per million. Consultation with the National Bureau of Standards on this point proved fruitless, but Taylor happened upon an advertisement from a small instrument company that had just the appropriate device under development, and it was adopted.

The results of this effort were announced in *Physical Review Letters* in February 1967, and full details were reported in *The Physical Review* in January 1969. The experiment actually addressed itself to two points: "First, can the frequency voltage ratio be measured sufficiently accurately to yield a value of e/h competitive with values obtained by other methods? Second, can the frequency-voltage ratio really be identified with $2e/h$? . . . The answer to the first question depends on establishing experimentally that there are no aspects of the ac Josephson effect itself which prohibit its practical application in this situation and also that absolute measurements of the frequency and voltage can be made with the required accuracy. The second question can be discussed either theoretically or experimentally, but the ultimate answer must rest on experimental evidence. In the absence of any theoretical evidence to the contrary, the Josephson frequency-voltage relation is presently thought to be theoretically exact. We have taken the view that a firmer foundation for a claimed determination of e/h can be laid by experimentally testing the invariance of the frequency-voltage ratio under a wide variety of experimental conditions. A demonstrated invariance at some level of precision constitutes strong circumstantial evidence that the ratio is equal to some fundamental physical quantity to that precision. The theoretical results can then be used with Occam's Razor[28] to identify the fundamental quantity with $2e/h$."

Space does not permit doing justice to the experimental details. Suffice it to say that the experiment gave a clear-cut affirmative answer to both questions. A variety of combinations of materials were used, at several levels of microwave power; and in addition to thin-film junctions, measurements were made on structures of a type called *point contacts*, described in the detailed report as "made by forming a sharp point on a superconducting wire . . . and pressing it against another superconductor. The contact pressure is quite critical, since it is the area of contact between the two superconductors which determines the strength of the coupling." The arrangement for use with such a device is shown in Fig. 14-16. All the results were in essential agreement, giving a value[29]

$$\nu/V \; = \; 483.5976 \text{ MHz/V}_{\text{NBS}} \; \pm \; 2.4 \text{ ppm};$$

the constancy formed what the authors regarded as "convincing evidence" on the second point. They could therefore deduce a result

$$h/e \; = \; 4.135 \; 707(15) \times 10^{-15} \text{ J sec/C},$$

which was indeed comparable with the values obtained by other methods.

This measurement and the technique whose power it demonstrated have had widespread consequences. For one thing, the method

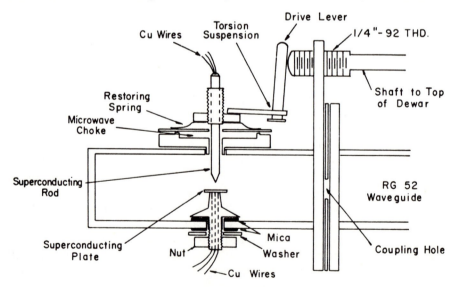

Fig. 14-16. Schematic diagram of the experimental arrangement for use of point contacts in the measurement of *e/h*. Note the device for adjusting the contact pressure. [*Phys. Rev.* **177** (1969), p. 646, Fig. 4.]

has been revealed as a workable means for intercomparison of voltages to the precision required by various national standards laboratories and will almost certainly supplement and perhaps supplant the clumsy and costly shipment of standard cells in a temperature-controlled enclosure. The International Bureau of Weights and Measures has already taken some beginning steps in this direction.

More fascinating, however, is the fact that this measurement, involving a phenomenon of solid-state physics, inherently a many-body effect, with a characteristic energy of a few microvolts, has repercussions in many other fields, particularly high-energy, elementary-particle physics. The reason for this is that the measurement makes possible a new computation of the "best values" of the basic natural constants e, h, $\alpha \equiv e^2/\hbar c$, m_e (the mass of the electron), and N_0 (Avogadro's number). Previous computations had all required at least one datum whose deduction from raw experimental results involved the use of quantum electrodynamics, the theory of the interaction of charged particles with electromagnetic radiation; and the value of α itself plays an important part in that theory. Despite this partially circular reasoning, the resulting value of α led to some inconsistencies in quantum electrodynamics. For example, the experimental datum used was the fine-structure splitting in deuterium; then calculation of the hyperfine structure of hydrogen gave a result that could be reconciled with experiment only by ascribing to the "excited states" of the proton effects much stronger than appear reasonable. If the deuterium fine-structure splitting was replaced as a basic datum by the Josephson-effect value of e/h, the resulting value of α led to a hydrogen hyperfine structure in agreement with experiment.[30]

Thus the Josephson effect has had a wide variety of consequences—and there is no reason to suspect that its possibilities have yet been exhausted.

Notes

1. Bardeen thus became the first person to win two Nobel prizes in the same field and only the third to win two at all.

2. The phenomenon is a consequence of the wavelike properties of particles (see *Crucial Experiments*, chap. 10), and particularly of the fact that the wave function does not vanish completely in regions where the classical kinetic energy would be negative. The effect has a complete parallel in connection with total internal reflection of electromagnetic waves; a good example is the microwave case described by J.J. Brady, R.O. Brick, and M.D. Pearson, *J. Opt. Soc. Amer.* **50**, 1080 (1960). It can also be demonstrated with water waves in a ripple tank.

3. See also the discussion of early theories of rectifier action in Chap. 9.

4. The paper presents the methods first and the concepts later.

5. What is altered is the relative positions of the two distributions. If the potential difference is V, the entire distribution for the metal at the higher potential is lowered (electrons are negative!) by an amount eV. The succeeding argument will be easier to follow if the reader can visualize this displacement.

6. Only one other group, at Arthur D. Little, Inc., was involved in tunneling studies before Giaever's success.

7. This makes possible the control of the temperature, since the boiling point of the liquid helium decreases as the pressure is reduced; cf. Chap. 5.

8. See chap. 5, p. 71.

9. This is the portion of the current-voltage characteristic in part (c) of Fig. 14-1 in which the current decreases with increasing voltage.

10. Although Fig. 14-2 is not necessarily drawn to scale, these dimensions provide a feel for the size of the apparatus.

11. The wave function for a particle in a classically forbidden region depends on the distance from the boundary of the region roughly according to the functional form $e^{-x/l}$; the "constant" l is the decay length.

12. It has the value $(\pi/eAR_{NN}) \Delta_1\Delta_2/(\Delta_1 + \Delta_2)$, where Δ_1 and Δ_2 are half the BCS energy-gap values for the respective superconductors, R_{NN} is the (tunneling) resistance of the junction when both metals are in the normal state, A is the area of the junction, and e is the electronic charge.

13. It was Anderson who noted that there would be a strong effect produced by a magnetic field.

14. The dashed line at higher voltages represents a jump across the negative-resistance region of the I-V curve.

15. This term refers to the fact, predicted in 1950 by Fritz London, that the magnetic flux through the hole in a multiply connected superconductor ("doughnut" or hollow cylinder) cannot have arbitrary values but is quantized. London's original deduction, based on his phenomenological theory together with very general properties of electron wave functions, gave the value hc/e for the quantum unit; the success of the BCS theory, showing that correlated pairs were the essential elements in superconduction, implied that the factor e in the denominator should be replaced by $2e$. The theory, including the revised value of the quantum of flux, was confirmed in two independent (and somewhat different) experiments in 1961: one by Bascom S. Deaver and William M. Fairbank at Stanford University and one by R. Doll and M. Näbauer at the Bavarian Academy of Sciences. The value of the flux quantum $hc/2e$ is 2.1×10^{-7} gauss $\cdot$ cm^2.

16. The power generated due to thermal noise in a circuit element is proportional to the resistance of the element.

17. Nicol, Shapiro, and Smith had an example in their Letter on measurement of the BCS gap; so did Giaever in his; and Giaever and Megerle had one in their long paper. But it was easy to conceive that a flaw in the oxide film had permitted the development of a superconductive short circuit. Not that there were not arguments to the contrary. Giaever, in discussing the example that occurs in the long paper, states his recollection that "a *surprisingly* small magnetic field made the 'short' disappear, i.e., consistent with the Josephson effect." Moreover, he knew of the work of Holm and Walther Meissner and of Dietrich, who found that insulating films between superconductors would carry

currents at zero voltage and convinced themselves that superconducting short circuits were *not* responsible; and he may have known of similar results obtained by Hans Meissner. However, he states, they were "too busy to reflect upon this [magnetic field effect] at the time." He himself could not think of a way to distinguish between a short circuit and the dc effect—Josephson's paper had been published by that time, but experimenters seem to have found it hard to understand—and he was involved in writing a Ph.D. thesis. He suggests that perhaps this should serve as a lesson against "having a too closed mind." In all fairness, however, note must be taken of a remark made by Anderson in the account quoted earlier, that as "a result of our contact with Josephson," he and Rowell "knew what to look for ... [and] understood what we saw." The experimenters were not the only shortsighted ones, either: Josephson had regarded his calculation as impossibly difficult until Anderson showed him a preprint of a calculation by M.H. Cohen, L.M. Falicov, and J.C. Phillips confirming Giaever's "naive" formula by a very simple method—and dropping the pair-current term!

18. They themselves had originally been convinced by a much more qualitative procedure, described by Rowell this way: "We checked the magnetic field sensitivity by moving a hefty bar magnet near the dewar. This caused the x-y recorder to jump abruptly from the zero voltage to finite voltage states and, in order to see just how sensitive the effect was, Phil moved down the laboratory until he was standing near the door, still swinging the bar magnet in his hand while the x-y recorder continued to jump erratically between zero and finite voltage as the field was varied by the distance and the angle of the magnet."

19. The "I" stands for insulator, not iodine.

20. The terms with $n = 0$ and $n = \pm 1$ can be obtained approximately in a straightforward way. With the abbreviation $\omega_J = 2eV_0/\hbar$, Eq. (14-2) is $j = j_1 \sin [\omega_J t + (2eV_1/\hbar\omega) \sin \omega t + \delta_0]$. Use of the standard formula for the sine of the sum of two angles gives

$$j = j_1 \{\sin(\omega_J t + \delta_0) \cos[(2eV_1/\hbar\omega) \sin \omega t]$$
$$+ \cos(\omega_J t + \delta_0) \sin[(2eV_1/\hbar\omega) \sin \omega t] \}.$$

Now assume that $2eV_1/\hbar\omega$ is small (i.e., much less than 1); then $\sin [(2eV_1/\hbar\omega) \times \sin \omega t] \cong (2eV_1/\hbar\omega) \sin \omega t$, and $\cos [(2eV_1/\hbar\omega) \sin \omega t] \cong 1$, which gives $j \cong j_1 [\sin (\omega_J t + \delta_0) + (2eV_1/\hbar\omega) \sin \omega t \cos (\omega_J t + \delta_0)]$. The second term in the brackets can be rewritten by use of another standard trigonometric identity, with the result

$$j \cong j_1 \{\sin(\omega_J t + \delta_0) + (eV_1/\hbar\omega) \sin[(\omega_J + \omega)t + \delta_0]$$
$$- (eV_1/\hbar\omega) \sin[(\omega_J - \omega)t - \delta_0]\}.$$

21. Note that in Figs. 14-8 and 14-9, current is plotted horizontally and voltage vertically, in contrast to the arrangement in Figs. 14-1, 14-5, 14-6, and 14-11.

22. Shapiro was able to make use of the fact that his resonant cavity had a mode within the range of an available 25-GHz (K-band) klystron, because of the fact that the standard wave guide used in the 9-GHz range (X-band) had a height that was nearly the same as the width of the standard K-band guide. Thus a suitable transition section would permit only the fundamental K-band mode to propagate into the X-band wave guide.

23. For a barrier between two different superconductors, the voltage values as given by Dayem and Martin are $(1/e)$ $(\epsilon_1 + \epsilon_2 \pm nh\nu)$, where ϵ is half the gap. The mechanism is that an electron which would normally be opposite the gap and thus unable to tunnel can change its energy, by absorption or emission of n photons, enough to be able to tunnel.

24. It was the existence of these steps that had been used earlier by Coon and Fiske as indirect evidence for the ac effect.

25. A significant part of the reason for their failure was the presence of television transmitters atop Mt. Wilson, just above the campus.

26. The plastic was dissolved in a volatile solvent and applied with a fine brush.

27. The enclosure is used for controlling the temperature of the cells.

28. Occam's, or Ockham's, Razor is the philosophical principle that the assumptions used to explain a thing should not be unnecessarily multiplied.

29. The subscript NBS means evaluation in terms of the unit of voltage maintained at the National Bureau of Standards. The value quoted refers to the unit used before 1 January 1969; a change made then for other reasons alters the value to 483.5935 MHz/V_{69NBS}.

30. This implies an error, as yet unidentified, in the deuterium value.

Bibliography

A general discussion of tunneling in studies of superconductors is given in Ivar Giaever and Karl Megerle, *The Physical Review* **122**, 1101 (1961); this paper contains references to earlier work. See also M. D. Fiske and I. Giaever, *Proceedings of the IEEE* **52**, 1155 (1964); R. W. Schmitt, *Physics Today* **14**, No. 12, 38 (1961).

Josephson's original predictions appear in B. D. Josephson, *Physics Letters* **1**, 251 (1962). This paper is very concise and abstract. A somewhat more discursive, though still fairly sophisticated, presentation appears in B. D. Josephson, *Advances in Physics* **14**, 419 (1965). See also Philip W. Anderson, *Physics Today* **23**, No. 11, 23 (1970); R. P. Feynman, R. B. Leighton, and M. Sands, *The Feynman Lectures on Physics*, vol. III (Addison-Wesley, Reading, Mass., 1965), chap. 21, especially secs. 21-29.

The initial reported observation of a Josephson effect, the "dc effect," is P. W. Anderson and J. M. Rowell, *Physical Review Letters* **10**, 230 (1963); the confirmatory evidence was given in J. M. Rowell, ibid. **11**, 200 (1963). The first observation of a form of the ac effect was reported by Sidney Shapiro, ibid. **11**, 80 (1963). Detection of the microwave radiation was reported by Ivar Giaever, ibid. **14**, 904 (1965). Actual coupling of the radiation to an external device was reported by I. K. Yanson, V. M. Svistunov, and I. M. Dmitrenko,

Zhurnal Eksperimental'noi i Teoreticheskoi Fiziki 48, 976 (1965) [translation appears in *Soviet Physics JETP* 21, 650 (1965)], and by D. N. Langenberg, D. J. Scalapino, B. N. Taylor, and R. G. Eck, *Physical Review Letters* 15, 294 (1965); see also Donald N. Langenberg, Douglas J. Scalapino, and Barry N. Taylor, *Scientific American* 214, No. 5, 30 (1966).

The "double-slit" interference was initially reported by R. C. Jaklevic, John Lambe, A. H. Silver, and J. E. Mercereau, *Physical Review Letters* 12, 159 (1964), and the effect of a vector potential in a free-field region by R. C. Jaklevic, J. J. Lambe, A. H. Silver, and J. E. Mercereau, ibid. 12, 274 (1964); this work is described in detail in R. C. Jaklevic, J. Lambe, J. E. Mercereau, and A. H. Silver, *The Physical Review* 140, A1628 (1965).

The preliminary experiment on e/h is reported in D. N. Langenberg, W. H. Parker, and B. N. Taylor, *The Physical Review* 150, 186 (1966). Results from the definitive experiment were presented in W. H. Parker, B. N. Taylor, and D. N. Langenberg, *Physical Review Letters* 18, 287 (1967), and the experiment was discussed in detail in W. H. Parker, D. N. Langenberg, A. Denenstein, and B. N. Taylor, *The Physical Review* 177, 639 (1969). The results are discussed in the context of a general evaluation of fundamental constants in B. N. Taylor, W. H. Parker, and D. N. Langenberg, *Reviews of Modern Physics* 41, 375 (1969); see also B. N. Taylor, D. N. Langenberg, and W. H. Parker, *Scientific American* 223, No. 4, 62 (1970); John Clarke, *American Journal of Physics* 38, 1071 (1970).

Some indications of still other applications of the effects are given in the Article by Anderson in *Physics Today* cited above; see also John Clarke, *Science* 184, 1235 (1974).

Chapter 15

Higher Symmetry for Elementary Particles

Before 1932, the only known entities that could be regarded as "elementary particles" were the proton and the electron. In that year the positron and the neutron were discovered; and since then the list has continued to grow, sometimes at a pace that seems almost alarming. The total number as of 1975 depends somewhat on what is to be considered as an established and distinct particle, but it is definitely over fifty and can run well over 100. Most of them are unstable, the exceptions being the photon, the electron, the proton, and two species of neutrino (one associated with the electron and the other with the muon).

Such an apparent variety among objects supposed to be elementary is rather paradoxical and suggests the possibility of close relationships among the various particles. In the search for these relationships, attention has primarily been focused on the *hadrons*, particles subject to the strong interactions, as the others (the leptons and the photon) are too few in number to be likely to reveal much.

One theory that provides simplifying relationships among the hadrons was proposed in 1961 by a theorist at the California Institute of Technology, Murray Gell-Mann, and independently by Yuval Ne'eman, then a colonel in the Israeli army serving as a military attaché in London and working at the Imperial College. At the time, there were no known data that contradicted the theory. As further data accumulated, the point was reached where the theory faced a crucial test: It predicted the existence of a previously unobserved particle, giving an approximate value for its mass and definite values for certain of its other properties. Experimentalists began trying to produce the particle, and in 1964 a group at Brookhaven National Laboratory reported success, in a Letter in *Physical Review Letters*. The evident usefulness of the theory was a factor in the awarding of a Nobel prize for physics to Gell-Mann in 1969. This chapter describes how the prediction came about and how the experiment was done.

There are at least two approaches to the problem of finding a scheme for organizing the elementary-particle states. One is to note that in atomic physics, the fundamental fact is the existence of a set of states for each atom, linked to one another by the emission or absorption of light and separated by energy intervals of the order of electron volts, a value governed by the strength of the electromagnetic interaction that produces atomic binding. Similarly, any nucleus has a set of excited states whose spacings, of the order of thousands of electron volts, are governed by the strength of the interaction between nucleons. Perhaps, then, the array of elementary particles is really the excited states of a much smaller set of entities, with the spacings—millions of electron volts—governed by some yet unknown interaction.[1] This view is strongly supported by some physicists (notably Victor Weisskopf); and, as will be discussed at the end of the chapter, there are some indications of such structure. As yet, however, the evidence is too sparse to be conclusive.

There is another approach, the principles of which can also be understood by an example from atomic physics. Consider the lowest few energy levels of the sodium atom, as shown in Fig. 15-1. It is to be noted that except for the first column, the levels occur in pairs having identical values of L but different values of J. Similar diagrams can be drawn for many other atoms, revealing not only doublets but higher multiplets. The explanation is well known: These structures are caused by the interaction between the intrinsic magnetic moment of the electron and the equivalent current resulting from the "orbital motion" of the electron. The point is that there is an approximate symmetry involved. If there were no magnetic moment associated with the electron spin, the energy would be insensitive to the orientation of the spin, and the multiplets would reduce to single energy values. The existence of the magnetic moment removes this symmetry, making the energy depend on the relative orientation of the vectors **L** (total orbital angular momentum) and **S** (total spin); the form of the additional term is $f(r)(L_x S_x + L_y S_y + L_z S_z)$, and from this fact alone can be derived several quantitative features such as relative spacings between members of multiplets (except doublets) and intensity ratios of transitions be-

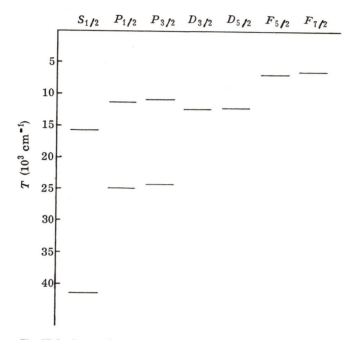

Fig. 15-1. Some of the energy levels of the sodium atom. The labels at the top indicate the orbital angular momentum L according to the scheme given in note 1 of Chap. 7, together with the total angular momentum J as a subscript. The vertical scale gives the so-called term values, which are proportional to the binding energies (10^3 cm^{-1} corresponds to 0.124 eV).

tween multiplets. The idea, then, is to find groupings of this kind among elementary particles and to account for them in mathematical terms from which analogous quantitative deductions can be made. It was this approach that was used by Gell-Mann and Ne'eman.

Their scheme was not the first successful use of symmetry ideas in dealing with elementary particles; in fact, part of its strength lay in the fact that it incorporated the most significant (and the oldest) of prior schemes of the sort, put forward in 1936 by Benedict Cassen and E. U. Condon. Cassen and Condon noted that such differences as exist in the forces between neutron-neutron, proton-proton, and neutron-proton pairs (in like spin states—nuclear forces were already known to be spin-dependent) could be almost entirely understood as due to the electrical charge on the proton. They therefore adopted a suggestion made by Werner Heisenberg in 1932, that the proton and neutron could be treated as merely two charge states of a single heavy particle. (The single particle has since become known as the *nucleon.*) Calculations of purely nuclear effects should then be, to a first approximation, independent of whether any given one of the particles involved was a proton, a neutron, or a linear combination of the two—the last possibility being of course nonphysical but readily expressible in mathematical terms. In other words, nuclear theory had to be approximately insensitive to linear transformations, arbitrary within certain restrictions, of the two elements "proton" and "neutron."

The mathematical apparatus needed to describe this insensitivity is completely parallel to that used ten years earlier to deal with electron spin; for this reason the variable whose value distinguishes proton from neutron became known as isotopic spin, isobaric spin, or (most simply and perhaps least misleadingly) isospin. Just as the mathematical formalism of spin allows not only spin $\frac{1}{2}$ but any integer or half-odd-integer value, so that of isospin allows not only particles with two charge states but particles with any integral number. This proved useful in developing the consequences of the 1935 proposal by Hideki Yukawa that nuclear forces are mediated by the emission and absorption of mesons by nucleons, as the theory called for positive, negative, and neutral mesons treated on equal footing and thus also capable of being regarded as states of a single particle. The continued parallel with the theory of spin led to a visualization of the isospin formalism as describing invariance under "rotations" in a three-dimensional abstract "space." The breaking of the symmetry by the effects of electric charge corresponds to singling out a particular "direction" in that space.

The years around 1950 witnessed the discovery of a new crop of particles, first observed in cosmic-ray events and soon thereafter produced by the beams from large accelerators. Many of them quickly became known as "strange" particles because of a peculiarity in their behavior: Their production was so copious that they had to be assumed to be hadrons, but their decays were so slow that they could be due only to the weak interaction of β decay (see Chap. 11). The problem was to understand why a hadron did not decay through the strong interaction. The solution, put forward independently by Gell-Mann and Kazuhiko Nishijima, was the existence of another approximate symmetry, manifesting itself in a new quantum number (called, naturally enough, *strangeness*). The symmetry held, and the quantum number was conserved, in the strong interactions, with the particles being produced in groups—usually pairs—of total strangeness zero. But their masses were such that they could not decay by modes that would conserve strangeness,[2] but only by the weak interaction for which the new symmetry does not hold. It was quickly recognized that a particle's strangeness S, its baryon number B, its component I_3 of isospin in the "special" direction mentioned earlier, and its charge Q were related by

$$Q = I_3 + \tfrac{1}{2}(B + S).$$

By the end of 1960, there were eight well-established meta-stable[3] baryon states and seven meson states, together with strong indications, either theoretical or experimental, of several additional meson states (some of which were unstable). They are displayed in Fig. 15-2. Among the baryons, at least, there is a suggestion of a grouping analogous to that of the multiplets in atomic energy levels: several states, all having some quantum numbers in common, with energy differences small compared to their separation from other such groups. The situation regarding the mesons is less clear-cut, but here too there are possible groupings, which might well be a consequence of the same symmetry as governed the baryons—if there really was one. It would evidently be only an approximate symmetry, and a suitable choice of the form of its breaking would provide relationships among the masses of the particles within a group.

Such a symmetry scheme was the one proposed by Gell-Mann and Ne'eman. It utilized a formalism which, though expressible only in terms of rather abstract mathematics, was actually the simplest generalization of the isospin formalism. The scheme leads straightforwardly to a grouping of particles into sets called *supermultiplets*.[4] All the particles in one supermultiplet would have the same spin and parity; but each supermultiplet would have a characteristic pattern of different values of isospin and hypercharge (hypercharge Y is the sum of baryon number and strangeness). Particularly prominent were supermultiplets of eight members, just such as to fit the eight baryons of Fig. 15-2. The seven spin-0 mesons also fit into such a pattern, with room for an eighth—an isospin singlet of hypercharge 0. The four proposed spin-1 meson states also fitted into such a scheme, with room for four others—two isospin doublets of hypercharge ± 1—for which role the K^* was a likely candidate. Moreover, a simple assumption about the breaking of the symmetry led to a relation among the baryon masses,

$$\tfrac{1}{2}(M_N + M_\Xi) = \tfrac{3}{4}M_\Lambda + \tfrac{1}{4}M_\Sigma,$$

where N stands for nucleon and where average values of the masses are to be used; this relation was quite well satisfied. A similar relation was predicted for the squares of the masses of the mesons, but it could not be checked, as one of the members of the set was not yet known. Partly because of the special importance of these sets of eight, the scheme became known as the "eightfold way," a phrase borrowed from Buddhism.

An eighth spin-0 meson, now known as the η, was discovered in

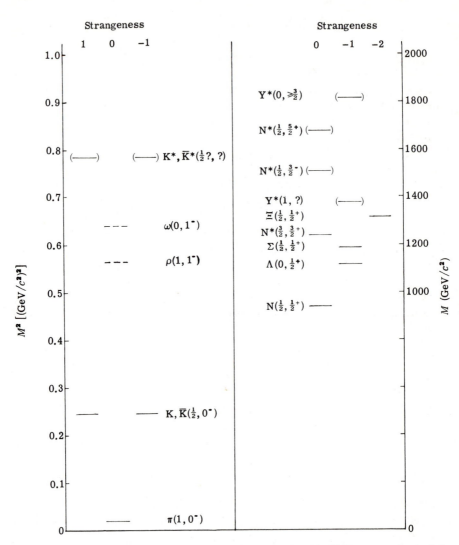

Fig. 15-2. The particle states known or suspected in 1960: mesons on the left, baryons on the right. The vertical scale is mass for the baryons, and square of mass, which for theoretical reasons is more appropriate for particles of integer spin, for the mesons. Particles in the same vertical column have the same strangeness. The symbol for each particle is followed in parentheses by its isospin *I* (the number of charge states is 2*I* + 1), its spin *J*, and its intrinsic parity. The states shown in parentheses had some experimental foundation but were not yet firmly established; those shown as dashed lines were only theoretical predictions.

1962; its mass is about 550 MeV/c^2, compared with the theoretical prediction of 563 MeV/c^2—not a serious discrepancy. The case of the spin-1 mesons is a bit more complex. When the existence of the ω was confirmed experimentally in 1961, its mass was found to be 790 MeV/c^2, whereas the mass formula predicted a value in the neighborhood of 900 MeV/c^2. Gell-Mann noted that the theory also could accommodate a ninth spin-1 meson, not grouped with any other particles, having the same isospin and hypercharge as the ω; if in the absence of symmetry-breaking interactions the mass of this ninth particle were close to 900 MeV/c^2, then it and the ω would interact or "mix" so as to yield two observable states whose masses would be roughly equally far from the unmixed value, in opposite directions. He therefore suggested that there might be another spin-1 meson at about 1000 MeV/c^2.

As of mid-1962, then, the scheme worked quite well for the baryons and the spin-0 mesons, but not so well for the spin-1 mesons. There was also another area of uncertainty. This had to do with the existence of fairly well-defined states that are unstable under the strong interactions and thus decay too rapidly to be observed as particles, but appear as resonances in meson-baryon scattering. One set of four of these had been known since the early 1950's; designated $N^*(1238)$, it was an isospin quartet with a mass of 1238 MeV/c^2, spin $3/2$, and strangeness 0. Others were being discovered as experiments at higher energies were performed. The question was whether the symmetry scheme could deal with them as well.

The mathematical aspect was straightforward. Each meson and each baryon was a member of an octet, and the scheme gave definite rules for determining what supermultiplets could result from a combination of two octets: There could be a singlet, two new octets, two groups of ten, and a group of 27. The scheme also specified what combinations of isospin and hypercharge could occur in each group. It could not, however, predict which of these supermultiplets would actually be realized in nature. Mass relations were no use, for while the theory provided mass relations within a supermultiplet, it could not give relations between supermultiplets, as these depend on dynamical details of the strong interactions. The $N^*(1238)$ gave no clue, as it could fit either into one of the decuplets or into the 27-plet.

At an international conference on high-energy physics held in July 1962 at CERN, the European Center for Nuclear Research in Geneva, Switzerland, three pieces of evidence were reported that bore on the problem. One was the existence of a spin-1 meson of hypercharge 0 and isospin 0, with a mass of 1020 MeV/c^2, about where Gell-Mann had predicted. This resolved the discrepancy of the spin-1 mesons and bolstered confidence in the essential correctness of the scheme. The second was the existence of a new baryon-meson resonance, an isospin doublet called the Ξ^*, at 1535 MeV/c^2. The third was the absence of any resonance structure in K^+-p scattering for energies from 140 to 800 MeV. Gell-Mann, in the discussion of this last report, noted that if the $N^*(1238)$ belonged to the 27-plet, then there should be such structure, as the 27-plet also included states with isospin 1 and hypercharge 2, implying strangeness 1 and charge up to +2, the values for the K^+-p system. On the other hand, the decuplet would contain an isospin triplet of strangeness −1, an isospin doublet of strangeness −2, and an isospin singlet of strangeness −3, with the masses of the multiplets equally spaced. A candidate for the triplet was already known, the Y_1^* with a mass of 1385 MeV/c^2 and spin probably 3/2; the newly discovered Ξ^* might be the doublet, since its mass was about the right value. This then implied the existence of a new particle, which Gell-Mann dubbed the Ω^-, at a mass of about 1685 MeV/c^2. This mass is low enough so that no strong decay, in which strangeness has to be conserved, is energetically possible, and thus the particle would be metastable. Gell-Mann suggested that it might be produced in the reaction

$$K^- + p \rightarrow K^+ + \bar{K}^0 + \Omega^-,$$

with incident K^- particles of momentum at least 3.5 GeV/c.

Among his hearers was Nicholas Samios of Brookhaven National Laboratory, who had previously reported the finding of the ninth spin-1 meson and the $\Xi^*(1535)$. Samios realized that Brookhaven was well equipped to search for the proposed particle. A huge bubble chamber, 80 inches long, was already being built and would serve as an excellent detector; and a beam that could provide K^- mesons of appropriate momentum was already being designed for use with the bubble chamber. Accordingly, after his return to Brookhaven, Samios assembled a team (the published report had 33 authors) to perform the experiment.

In a bubble chamber, a liquid near its boiling point is suddenly brought into a superheated state by expansion which reduces the pressure and thereby lowers the boiling point. Vaporization begins along the trails of ions left by the passage of fast charged particles through the liquid, and the nascent vapor forms a string of bubbles that can be photographed. Normally at least two cameras are used (the 80-inch uses three), pointed at different angles to permit three-dimensional reconstruction of the tracks. Frequently the chamber is placed in a magnetic field which makes the charged particles travel in curved paths (the field used in the present experiment was about 20 kilogauss). Analysis of the properties of the tracks—curvature and bubble density—then yields the momenta and identities of the initial particles. The bubble chamber is thus closely analogous to the much older cloud chamber, where the trails were droplets condensed from a supersaturated vapor onto the ions formed by passage of a charged particle. It has a great advantage in that the density of a liquid is much higher than that of a gas, so that a particle of a given energy does not have to travel so far before being brought to rest, and a photon is more likely to convert into electrons. Also, it can be recycled somewhat more rapidly than the cloud chamber can.

Many different liquids can be used in bubble chambers. The 80-inch chamber uses hydrogen. Compared to other liquids, hydrogen gives the smallest factor of advantage in density (though still far better than a vapor); on the other hand, it has the feature that there are no complex nuclei with which the particles of interest may collide to produce confusing interactions. From a more mundane point of view, its use poses considerable problems: It is highly dangerous with respect to fire and explosion, requiring extensive safety precautions; and its boiling point at ordinary pressure is about $-253°C$ (its critical temperature is $-241°C$), which implies the need for a large refrigerating capacity. A large hydrogen bubble chamber is therefore an imposing piece of machinery. A picture of the 80-inch chamber is shown in Fig. 15-3, and a cutaway drawing in Fig. 15-4.

The K^- mesons are produced by protons that have been accelerated to energies of about 30 GeV in the large accelerator at Brookhaven known as the *alternating-gradient synchrotron* ("AGS" for short). The protons strike a tungsten target at the end of one of the straight sections of the "doughnut," producing a spray (mostly in the forward direction) of particles of various kinds with a substantial range of energies. A crucial step is separating the particles of the desired kind and energy from this array and transporting them as a focused and shaped beam to the location of the detector. The system used is based on two principles: (1) A charged particle traveling perpendicular to a uniform magnetic field moves along an arc of a circle whose radius is inversely proportional to the momentum of the particle; consequently, the particle is deflected by an amount depending only on its momentum. (2) A charged particle traveling through uniform electric and magnetic fields that are perpendicular to each other and to the (initial) direction of motion of the particle undergoes zero deflection if and only if its speed v is equal (in mks units) to E/B, the ratio of the electric field strength to the magnetic flux density. The actual design of a beam transport system, however, is a rather complex problem in ion optics, arising from the facts that any slit system for defining the initial paths gives a small but nonzero

Fig. 15-3. The 80-inch bubble chamber at Brookhaven National Laboratory. 1, copper coil and iron yoke of the magnet; 2, beam window of the vacuum chamber surrounding the bubble chamber proper (this vacuum chamber functions like the vacuum space between the walls of a Dewar flask); 3, vacuum pump system; 4, expansion system assembly; 5, liquid hydrogen refrigerator, located behind stairway and magnet; 6, illumination and camera system; 7, undercarriage for moving the entire 450-ton assembly. (Photograph courtesy of Brookhaven National Laboratory.)

range of directions and that the production of electric and magnetic fields over finite regions of space gives rise to fringing fields which themselves have non-negligible focusing or defocusing effects. The system actually used, shown schematically in Fig. 15-5, involved seven dipole magnets for deflecting the beam, fifteen quadripole and two sextupole magnets for focusing it, five slits for selecting the desired portion, and two crossed-field separators, plus seven additional magnets for shaping and positioning the final separated beam. It was some 450 feet long, and the magnets consumed nearly 5 megawatts of electric power. The whole beam path was inside an evacuated tube to keep the particles in the beam from being diverted by collisions with air molecules.

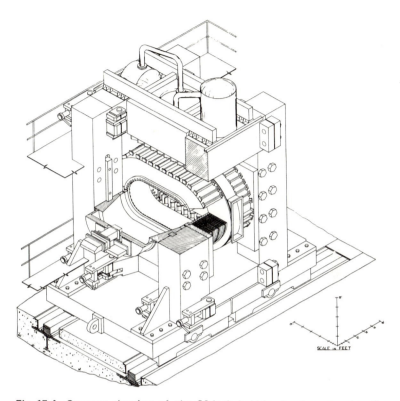

SCALE IN FEET

Fig. 15-4. Cutaway drawing of the 80-inch bubble chamber, showing the magnet coils and the chamber proper.

All did not go entirely smoothly. The beam transport system did not work properly at first, and it was not until September of 1963 that the first pictures from the chamber were produced. In addition, there were influential theorists who expressed considerable doubts as to the soundness of the predictions. Meanwhile, a competing group at CERN, the European nuclear research center, was having its troubles. In their case, it was the bubble chamber that was the source of the difficulty, and the complexities of operating a multi-national establishment kept another one from being used instead. The Brookhaven group persevered; exposures for the experiment itself began in December 1963, and in February 1964 the report of success was published.

"The BNL 80-in. hydrogen bubble chamber was exposed to a mass-separated beam of 5.0-BeV/c K^- mesons at the Brookhaven AGS. About 100 000 pictures were taken containing a total K^- track length of $\sim 10^6$ feet. These pictures have been partially analyzed to search for the more characteristic decay modes of the Ω^-.

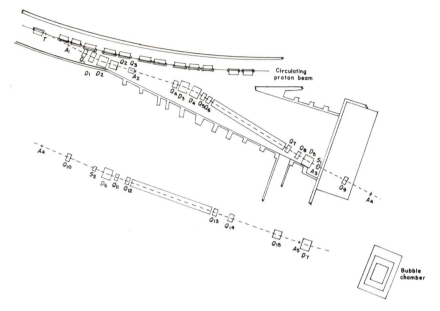

Fig. 15-5. Layout of the beam system for the Ω^- experiment, shown in two segments for space reasons. A portion of the AGS is at the upper left; the bubble chamber, at the lower right. T is the tungsten target; A_1 through A_5, slit apertures (A_3 is within S_1); D_1 through D_7, dipole magnets for beam steering and momentum dispersion; Q_1 through Q_{15} and S_1 through S_2, quadripole and sextupole magnets for focusing. The long rectangles without labels represent the crossed-field mass separators. Not shown are the final steering and beam-shaping magnets between D_7 and the bubble chamber. The dashed line shows the path of the desired beam.

"The event in question is shown in Fig. [15-6], and the pertinent measured quantities are given in Table [15-]I). Our interpretation of this event is

$$K^- + p \rightarrow \Omega^- + K^- + K^0$$
$$ \downarrow \Xi^0 + \pi^-$$
$$ \downarrow \Lambda^0 + \pi^0$$
$$ \downarrow \gamma_1 + \gamma_2 \qquad\qquad (\text{[15-]1})$$
$$ \downarrow e^+ + e^-$$
$$ \downarrow e^+ + e^-$$
$$ \downarrow \pi^0 + p.$$

From   the   momentum   and   gap   length   [between   bubbles]

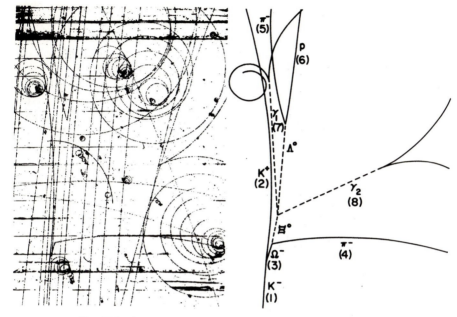

Fig. 15-6. "Photograph and line diagram of event showing decay of Ω^-." [*Phys. Rev. Lett.* **12** (1964), p. 205, Fig. 2.]

Table 15-I. Measured parameters of particle tracks in Fig. 15-6. [*Phys. Rev. Lett.* **12** (1964), p. 205, Table I.]

Track	Azimuth (deg)	Dip (deg)	Momentum (MeV/c)
1	4.2 ± 0.1	1.1 ± 0.1	4890 ± 100
2	6.9 ± 0.1	3.3 ± 0.1	501 ± 5.5
3	14.5 ± 0.5	−1.5 ± 0.6	. . .
4	79.5 ± 0.1	−2.7 ± 0.1	281 ± 6
5	344.5 ± 0.1	−12.0 ± 0.2	256 ± 3
6	9.6 ± 0.1	−2.5 ± 0.1	1500 ± 15
7	357.0 ± 0.3	3.9 ± 0.4	82 ± 2
8	63.3 ± 0.3	−2.4 ± 0.2	177 ± 2

measurements, track 2 is identified as a K^+. (A bubble density of 1.9 times minimum was expected for this track while the measured value was 1.7 ± 0.2.) Tracks 5 and 6 are in good agreement with the decay of a Λ^0, but the Λ^0 cannot come from the primary interaction." This fact, which is one of the keys to establishing the nature of the event, can be determined by kinematic analysis once the momenta and directions of the decay particles are measured; but Samios, who discovered the event himself, had used a much simpler method involving a trick known to those who work with such particle tracks: When two tracks from a decay are concave towards each other, as is the case with tracks 5 and 6, they may intersect again; then a straight line through that intersection and the decay point is the direction of motion of the invisible neutral parent. The intersection of tracks 5 and 6 is not shown on the published portion of the picture, but it was visible in the original pictures and provided one of Samios's clues. The report goes on: "The Λ^0 mass as calculated from the measured proton and π^- kinematic quantities is 1116 ± 2 MeV/c^2 [cf. the value given in Fig. 15-2]." Another conceivable possibility was that these tracks represented decay of a K^0; this was ruled unlikely because track 6 had a bubble density consistent with that for a proton but not for a π^+. "In any case, from kinematical considerations such a K^0 could not have come from the production vertex. The Λ^0 appears six decay lengths from the wall of the bubble chamber [making it unlikely that it originated in the wall], and there is no other visible origin in the chamber.

"The event is unusual in that two gamma rays, apparently associated with it, convert to electron-positron pairs in the liquid hydrogen." This was a lucky accident, and was discovered late in the analysis—after the identity of track 3 as the Ω^- was already fairly certain. "From measurements of the electron momenta and angles, we determine that the effective mass of the gamma rays is 135.1 ± 1.5 MeV/c^2, consistent with a π^0 decay. In a similar manner, we have used the calculated π^0 momentum and angles, and the values from the fitted Λ^0 to determine the mass of the neutral decaying hyperon to be 1316 ± 4 MeV/c^2 in excellent agreement with that of the Ξ^0. The projections of the lines of flight of the two gammas and the Λ^0 onto the XY plane (parallel to the film) intersect within 1 mm and in the XZ plane within 3 mm." Thus it is highly likely that the Λ^0 and the two gammas originate at essentially a single point; the distance that the π^0 could reasonably travel before decaying (its mean life is only about 10^{-15} sec) is negligible. "The calculated momentum vector of the Ξ^0 points back to the decay point of track 3 within 1 mm and misses the production vertex by 5 mm in the XY plane." Thus the Ξ^0 probably came from the decay of particle 3 and was unlikely to have been produced in the initial interaction. "The transverse momenta [that is, the components perpendicular to the direction of motion of the parent particle] of the Ξ^0 and track 4 balance within the errors, indicating that no other particle is emitted in the decay of particle 3.

"We will now discuss the decay of particle 3. From the momentum and gap length measurements on track 4, we conclude that its mass is less than that of a K. Using the Ξ^0 momentum and assuming particle 4 to be a π^-, the mass of particle 3 is computed to be 1686 ± 12 MeV/c^2 and its momentum to be 2015 ± 20 MeV/c^2. Note that the measured transverse momentum of particle 4, 248 ± 5 MeV/c^2, is greater than the maximum momentum for the possible decay modes of the known particles . . ., except for $\Xi^- \to e^- + n + \nu$. We reject this hypothesis, not only because it involves $\Delta S = 2$ [no process with $|\Delta S| > 1$ has even yet been observed], but also because it disregards the previously established associations of the Λ and two gammas with the event." It was this large transverse momentum of particle 4 that was the second key factor in establishing the nature of the event.

"The proper lifetime of particle 3 was calculated to be 0.7 × 10^{-10} sec; consequently we may assume that it decayed by a weak interaction with $\Delta S = 1$ into a system with strangeness minus two. Since a particle with $S = -1$ would decay very rapidly into $Y + \pi$ [where Y might be Λ, Σ, or any of several resonances], we conclude that particle 3 has strangeness minus three. The missing mass at the production vertex is calculated to be 500 ± 25 MeV/c^2, in good agreement with the K^0 assumed in Reaction ([15-]1). . . .

"In view of the properties of charge ($Q = -1$), strangeness ($S = -3$) and mass ($M = 1686 \pm 12$ MeV/c^2) established for particle 3, we feel justified in identifying it with the sought-for Ω^-."

The Brookhaven group soon found another example of an Ω^-, decaying by a different mode; and in ten years since, 39 more have been seen. The identity of the particle is well established. In addition, an analysis by Luis Alvarez in 1973 indicates that three cosmic-ray events seen in 1954 and 1955, and never previously interpreted in an entirely satisfactory way, are actually examples of Ω^- particles. Rather remarkably, each of the three has an anomalous aspect to its behavior: One undergoes interaction with a silver nucleus before decaying; one is captured into a pseudoatomic state; and in the third, the Λ from the decay $\Omega^- \to K^- + \Lambda$ is captured by a nitrogen nucleus by the reaction $\Lambda + N^{14} \to {}_\Lambda C^{13} + p + n$, the hypernucleus then decaying as ${}_\Lambda C^{13} \to C^{12} + \pi^0 + n$.

The story is not yet complete, however. For one thing, the scheme of Gell-Mann and Ne'eman requires that all states in a super-multiplet have the same spin. The spins of the $N^*(1238)$, the $Y_1^*(1385)$, and the $\Xi^*(1535)$ have all been established as 3/2; but that of the Ω^- is unknown. The determination of the spin of a particle is based on analysis of the angular distributions of its decay products, which requires several scores of examples of a single mode of decay. It is conceivable that the spin of the Ω^- may turn out to be different from 3/2, though no one takes the possibility very seriously.

Secondly, even the worth of the scheme as a method of classification is not entirely certain. As of mid-1974, three octets-cum-singlets of mesons and five octets and two decuplets of baryons have been grouped rather convincingly; the analysis takes into account not only mass relationships, but also the relative partial decay rates of particles of one group into those of another, which are likewise calculable from the theory. However, a large number of resonances, both meson and baryon, remain unclassified. Moreover, it is not at all understood why only singlets, octets, and decuplets have been observed from among the variety of multiplets that the theory allows.

A third problem, connected with the second, is the relationship between the symmetry scheme and other regularities in particle properties. For example, a widely studied formalism, known as the system of "Regge trajectories" and arising out of mathematical aspects of the theory of scattering, implies the existence of what might be called a "vertical" grouping of particles as contrasted with the "horizontal" grouping of the eightfold way. In this grouping, every particle should lie on a "ladder" of particles, all of the same intrinsic parity, of successively higher mass, and with spin increasing by two units at each step; the particles on a given ladder are referred to as "Regge recurrences" of the lowest member. This system has strong resemblances to the concept of "excited states" mentioned early in the chapter. Presumably, the Regge recurrences of any supermultiplet should themselves form supermultiplets of the same kind. But the size of the mass steps is not determined, nor are the spacings within a supermultiplet necessarily the same from one recurrence to the next. As yet, despite the existence of a plethora of baryon resonances, only a few Regge-recurrence associations appear firm, and these few do not include those of any complete supermultiplets.

Finally, there is a puzzling feature within the theory. The formalisms of both isospin and the eightfold way are based on the algebra of *matrices*, rectangular (in these cases, square) arrays of numbers that can be manipulated algebraically according to definite rules. Isospin, it is to be recalled, originated in the consideration of a restricted class of transformations between two states; consequently, the basic matrices are 2×2 matrices that satisfy certain conditions. The basic matrices of the eightfold way are 3×3 and subject to the same sort of conditions. Why, then, is there not a set of three entities whose conceptual transformations give rise to the 3×3 matrices? Gell-Mann has suggested that perhaps there actually are. He proposes that they would have spin 1/2; two would be nonstrange and one would have strangeness −1; and, in his proposal, their charges and baryon numbers would be one-third integral. He calls them "quarks." They have not yet been observed; presumably, if they exist, they have very large masses. The idea has some attractive features but also some severe drawbacks.

The case for higher symmetry, then, is still a field for research. It has proved to be a useful guide, but its validity is not certain.

Notes

1. It may be significant that the symmetry scheme to be discussed in this chapter suggests that the observed elementary particles may be bound systems of a set of three kinds of hypothetical particles called "quarks"; if they are, the binding forces between quarks must be far stronger than those between nucleons.

2. There is one exception: The Σ^0 decays by the electromagnetic interaction, $\Sigma^0 \rightarrow \Lambda^0 + \gamma$.

3. This means stable under strong interactions.

4. This term was introduced by E. P. Wigner, who applied the combination of spin invariance and isospin invariance—both only approximate symmetries—to the theory of nuclear structure.

Bibliography

The original proposals by M. Gell-Mann [*The Physical Review* **125**, 1067 (1962)] and Y. Ne'eman [*Nuclear Physics* **26**, 222 (1961)] are couched in abstruse mathematical terms. A sketch of the theory is contained in Geoffrey F. Chew, Murray Gell-Mann, and Arthur H. Rosenfeld, *Scientific American* **210**, No. 2, 74 (1964).

The experimental report is V. E. Barnes et al., *Physical Review Letters* **12**, 204 (1964). See also W. B. Fowler and N. P. Samios, *Scientific American* **211**, No. 4, 36 (1964).

The experimental status of the theory as of late 1973 is given in N. P. Samios, M. Goldberg, and B. T. Meadows, *Reviews of Modern Physics* **46**, 49 (1974).

Chapter 16

A Possible
Cosmological Clue

The process of discovery is not necessarily a chance affair. For example, the discovery of a cure for a specific disease is likely to be the result of a careful study of how the causative mechanism operates, so that a means of interfering with that operation can be effected. Nevertheless, chance enters frequently enough that the word "discovery" often carries the connotation of accident.

Naturally, the role of chance will be different in different fields. In astronomy, and especially in those aspects of astronomy that pertain to cosmology, it can be expected to be large, as our knowledge of the detailed structure of the universe is simply too limited for us to know where or how to look for a new phenomenon. Indeed, the whole subfield of radio astronomy arose by accident, when Karl Jansky of Bell Telephone Laboratories sought for the cause of a particular kind of radio static and found that it consisted of noise being received from near the direction of the center of our galaxy. From that beginning, radio astronomy has grown into a well-developed field and has been the means of a number of discoveries, some of which are not yet fully understood.

Among these discoveries, the one that has perhaps the greatest implication for cosmology is the observation of a background of microwave radiation, apparently universal in distribution and well represented (as far as accurate measurements have extended) as the spectrum of blackbody radiation at a temperature of about 3 K. This discovery, made in 1965 by A. A. Penzias and R. W. Wilson and reported in a Letter in the *Astrophysical Journal*, is described in this chapter.

First it will be helpful to discuss its significance. Among the theories of the development of the universe, one class is characterized by the assumption that the universe started in a spatially rather small, highly condensed state; the initial stages, at least, of the development from this state were so rapid as to be explosive, so that the theory is often referred to as the "big bang" theory. It was recognized as early as 1946 by George Gamow and his coworkers that the early stages of such a universe would be dominated by blackbody radiation, and that a remnant of that radiation should still be present. The expansion of the universe would have reduced its temperature to a value in the neighborhood of 5 K. The verification of the existence of such a remnant would thus give strong support to the big bang theory.[1]

The basis of the measurement is the fact that any electrical circuit element at a temperature above absolute zero generates noise as a consequence of thermal motions (principally of the conduction electrons). The noise power per unit bandwidth, that is, per unit frequency range, is directly proportional to the temperature, with a coefficient that depends in a known way on the electrical properties of the element. This relationship can be used backward to assign an effective noise temperature to any noisy element, regardless of whether or not the noise is of thermal origin. Thus Penzias and Wilson refer to[2] "measurements of the *effective* zenith noise temperature of the ... antenna." What they found was that there was a contribution to the temperature apparently due to radiation received from interstellar space.

The antenna used was a 20-foot horn-reflector type that had been used in connection with the Echo satellite program, and was described in 1961 by A. B. Crawford, D. C. Hogg, and L. E. Hunt.[3]

"The horn-reflector type of antenna ... is a combination of a square electromagnetic horn and a reflector that is a sector of a paraboloid of revolution, as illustrated in Fig. [16-]1. The apex of the horn coincides with the focus of the paraboloid. Since the antenna design is based on geometrical optics and has no frequency-sensitive elements, it is extremely broadband; it is not polarization-sensitive and can be used in any linear or circular polarization. ... Due to the shielding effect of the horn, the far side and back lobes[4] are very small.

"... [T]he low side and back lobes insure that when the antenna beam is pointed to the sky very little noise power is received from the ground; the antenna is thus a low-noise transducer which permits exploitation of the low-noise features of the maser amplifier.

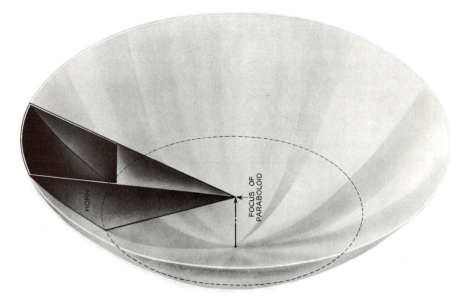

Fig. 16-1. "Sketch showing relationship of horn reflector antenna to a parabo-loid of revolution." [*Bell Syst. Tech. J.* **40** (1961), p. 1096, Fig. 1. Copyright 1961, American Telephone and Telegraph Company; reprinted by permission.]

"Fig. [16-]2 is a photograph of the horn-reflector antenna erected on the Crawford Hill site of the Holmdel Laboratory. . . . To permit the antenna beam to be directed to any part of the sky, the antenna is mounted with the axis of the horn horizontal. Rotation about this axis affords tracking in elevation while the entire assembly is rotated about a vertical axis for tracking in azimuth. The antenna is about 50 feet in length, the radiating aperture is approximately 20 by 20 feet, and the weight is about 18 tons. The structure is designed to survive winds of 100 miles per hour.

"The elevation structure, both horn and reflector, is con-structed of aluminum. The elevation wheel, 30 feet in diameter, supports all radial loads and rotates on rollers mounted on the base frame. All axial or thrust loads are taken by a large ball bearing at the apex end of the horn. The horn proper continues through this bearing into the equipment cab. Here is located a tapered transition section from square to round waveguide, a rotating joint, and waveguide takeoffs. . . . The ability to locate the receiver equipment at the apex of the horn, thus eliminating the loss and noise contribution of a connecting line, is an important feature of this antenna.

"The triangular base frame is constructed of structural steel shapes. It rotates on wheels about a center pintle ball bearing on a track 30 feet in diameter. The track consists of stress-relieved, planed steel plates which were individually adjusted to produce a track flat to about 1/64 inch. The faces of the wheels are cone-shaped to minimize sliding friction. A tangential force of about 100 pounds is sufficient to start the antenna in motion. . . .

"The antenna is driven in azimuth and elevation by 10 H.P. direct-current servo gear-motors. Power is transmitted by sprockets (with teeth specially cut for rack operation) to roller chains which are fastened to the vertical wheel and to the plates forming the horizontal track. . . . The maximum speed of rotation in both azimuth and elevation is 5° per second; the maximum acceleration for both axes is 5° per second per second. Power for the drives is brought to the rotating structure through a slip-ring assembly inside the small plywood house located over the center bearing (Fig. [16-]2). All the electrical circuits needed for the operation of the antenna and the receiving equipment in the cab come through the slip-ring assembly.

"Positional information for the antenna is derived from data units driven by large (48-inch) accurately cut and accurately aligned gears located on the bearings at the apex of the horn and at the center of the base frame. The data units contain synchro trans-mitters."

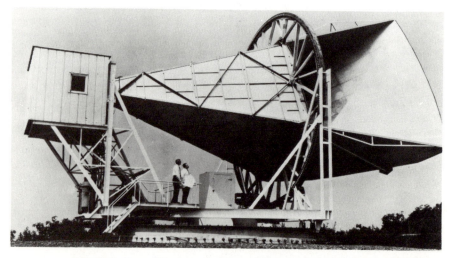

Fig. 16-2. Horn-reflector antenna used by Penzias and Wilson. [*Bell Syst. Tech. J.* **40** (1961), p. 1097, Fig. 2. Copyright 1961, American Telephone and Telegraph Company; reprinted by permission.]

The radiometer was described in detail by Penzias and Wilson in another paper reporting its use in another investigation.[5]

"[It] uses a traveling-wave ruby maser as the first stage [of amplification]. The maser . . . has a noise temperature of about $3.5°K$, a gain of 42 db,[6] and a band width of 15 mc/s. . . .

"The maser is connected to the horn-reflector antenna through the waveguide assembly shown in Figure [16-3]. The antenna feed is a round waveguide, while both the reference port and the output to the maser are rectangluar waveguides. The required transitions are effected by polarization couplers . . . [which] consist of a 'T' formed by a straight-through circular waveguide and a rectangular side arm. One of the couplers accepts two orthogonal . . . modes [of polarization], passes one straight through, and couples the other out the rectangular port. . . .

"The output of the maser is fed to a mixer, and the remainder of the radiometer is conventional."

Measurements were made by switching manually between the antenna input and a liquid-helium‒cooled reference termination. This termination, consisting essentially of a 4-foot-long piece of brass

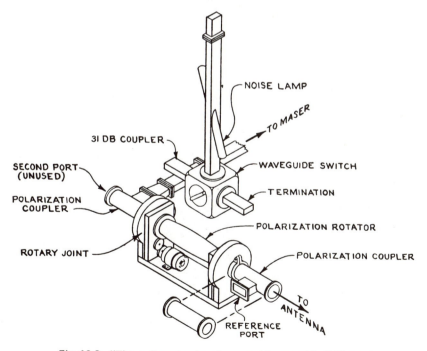

Fig. 16-3. "The radiometer input waveguide assembly." The noise lamp was used in calibration of the radiometer. [*Astrophys. J.* **142** (1965), p. 1150, Fig. 1. Copyright 1965 by the University of Chicago.]

waveguide terminated by an absorbing pyramid completely immersed in liquid helium, had been described earlier by Penzias. An isometric view is shown in Fig. 16-4.

"The waveguide is actually composed of two sections. The lower section contains a pyramid of absorbing material . . . and is completely filled with liquid helium during operation. The flange connecting the two sections holds a Mylar septum which excludes liquid helium from the upper section. This septum is set at an angle of 30° with the axis of the waveguide. The gas-liquid interface is thus wedge shaped when the lower waveguide is filled with liquid, and a smooth transition of guide impedance results. When the liquid helium level in the Dewar falls below the top of the lower flange, the gas-liquid interface is no longer tapered and good operation is no longer possible." The effective temperature of the device (approximately 5 K) was known to within 0.2 K.

Returning to the primary report, Penzias and Wilson state,[7] "The antenna, reference termination, and radiometer were well matched so that a round-trip return loss of more than 55 db existed throughout the measurement; thus errors in the measurement of the effective temperature due to impedance mismatch can be neglected. The estimated error in the measured value of the total antenna temperature is $0.3°K$ and comes largely from uncertainty in the absolute calibration of the reference termination."

The measurements gave a total antenna temperature, measured with the antenna pointed at the zenith, of 6.7 K. From this had to be subtracted a contribution due to atmospheric absorption, one due to Ohmic losses in the antenna, and one due to back-lobe response.

"The contribution to the antenna temperature due to atmospheric absorption was obtained by recording the variation in antenna temperature with elevation angle and employing the secant law.[8] The result, $2.3° + 0.3°$ K, is in good agreement with published values. . . .

"The contribution to the antenna temperature from ohmic losses is computed to be $0.8° + 0.4°$ K. In this calculation we have divided the antenna into three parts: (1) two non-uniform tapers approximately 1 m in total length which transform between the 2 1/8-inch round output waveguide and the 6-inch-square antenna throat opening; (2) a double-choke rotary joint located between these two tapers; (3) the antenna itself. Care was taken to clean and align joints between these parts so that they would not significantly increase the loss in the structure. Appropriate tests were made for leakage and loss in the rotary joint with negative results.

"The possibility of losses in the antenna horn due to imperfections in its seams was eliminated by means of a taping test. Taping all

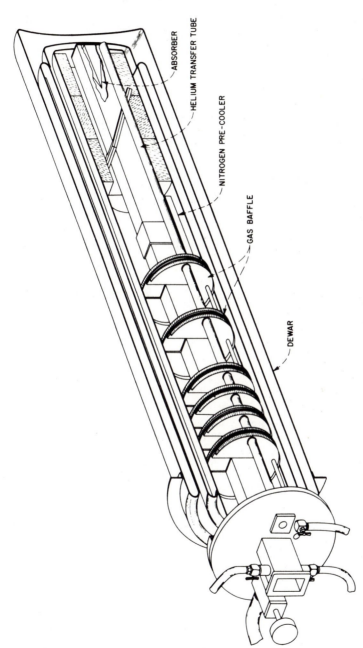

ABSORBER

HELIUM TRANSFER TUBE

NITROGEN PRE-COOLER

GAS BAFFLE

DEWAR

Fig. 16-4. Isometric view of the reference termination mounted in its Dewar. In use, the device is vertical with the closed end down. [*Rev. Sci. Instrum.* **36** (1965), p. 69, Fig. 1.]

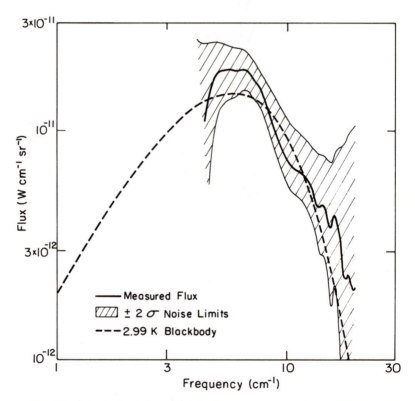

Fig. 16-5. Spectrum of the cosmic background as measured by Woody et al., compared with the blackbody spectrum for a temperature of 2.99 K. In the legend, σ is the symbol for "standard deviation." [*Phys. Rev. Lett.* **34** (1975), p. 1038, Fig. 2.]

the seams in the section near the throat and most of the others with aluminum tape caused no observable change in antenna temperature.

"The backlobe response to ground radiation is taken to be less than 0.1°K" as a result both of measurements on the antenna itself, using a small transmitter on the ground, and of more exact studies on smaller horn-reflector antennas which were expected to have poorer back-lobe characteristics.

"From a combination of the above, we compute the remaining

unaccounted-for antenna temperature to be $3.5° + 1.0°K$." The introductory paragraph had stated, "This excess temperature is, within the limits of our observations, isotropic, unpolarized,[9] and free from seasonal variations (July, 1964—April, 1965)." These were just the properties expected of a remnant blackbody radiation, and the temperature was acceptably near the calculated value.

For nearly a decade, while the validity of the results were unquestioned, it was uncertain whether the radiation was truly a blackbody remnant. Several other measurements made by other groups at wavelengths between 0.33 and 73.5 cm (Penzias and Wilson worked at 7.35 cm) were consistent with a blackbody spectrum at a temperature of about 2.8 K. However, the characteristic feature of the blackbody spectrum is that it passes through a maximum, which for that temperature occurs at a wavelength of about 0.11 cm. The question was whether the spectrum of the galactic radiation does indeed have such a maximum. The problem was that measurements at shorter wavelengths could not be made from the earth's surface because of atmospheric absorption. Some determinations were made from the absorption properties of interstellar molecules; but with one exception (the validity of which was questioned), these were only upper limits. Finally, in April 1975 there was published a measurement of the entire range from 0.33 to 0.025 cm, made with a balloon-borne detector by D. P. Woody, J. C. Mather, N. S. Nishioka, and P. L. Richards of the University of California and Lawrence Berkeley Laboratory. Their results are shown in Fig. 16-5. Although there is still a sizable uncertainty as to the exact curve, it is evident that the maximum exists. The authors state, "This measurement establishes that the cosmic background radiation has a thermal spectrum from 4 to 17 cm^{-1}," i.e., 0.25 to 0.059 cm.

It is still possible, of course, that the "big bang" theory will prove untenable. But any alternative will have to provide an explanation for the blackbody nature of the cosmic background radiation.

Notes

1. Penzias and Wilson, however, were unaware of Gamow's work until after their paper was published. Instead, they discussed their findings with R. H. Dicke, P. J. E. Peebles, P. G. Roll, and D. T. Wilkinson, who had independently repeated Gamow's deduction and who published their own work, modified somewhat in the light of the experimental results, in the same journal issue with Penzias and Wilson.

2. Quoted from *Astrophysical Journal*, vol. 142 (1965). Copyright 1965 by the University of Chicago.

3. Quoted from *Bell System Technical Journal*, vol. 40 (1961). Copyright 1961, American Telephone and Telegraph Company; reprinted by permission.

4. This means the pattern of response to radiation from the sides and behind the antenna.

5. See Note 2.

6. Relative power levels are commonly measured in *decibels* (abbreviated db or dB), a logarithmic ratio measurement. Specifically, the level difference in decibels is $10 \log_{10} (P_2/P_1)$, where P_1 and P_2 are the values of power (or energy, or intensity) being compared.

7. See Note 2.

8. The effect of the atmosphere is proportional to the path length through the atmosphere, which varies as the secant of the elevation angle.

9. The antenna itself is insensitive to polarization of the incident radiation, but different polarizations could be selected in the radiometer input.

Bibliography

The original report is A. A. Penzias and R. W. Wilson, *Astrophysical Journal* **142**, 419 (1965). The antenna is described in A. B. Crawford, D. C. Hogg, and L. E. Hunt, *Bell System Technical Journal* **40**, 1095 (1961); the radiometer, in A. A. Penzias and R. W. Wilson, *Astrophysical Journal* **142**, 1149 (1965); the reference termination, in A. A. Penzias, *Reviews of Scientific Instruments* **36**, 68 (1965).

Two papers review the state of affairs as of 1972; one emphasizes the overall situation and one concentrates on the results from interstellar molecules. The former is A. A. Penzias, in *Cosmology, Fusion, and Other Matters*, edited by F. Reines (University of Colorado Press, Boulder, Colorado, 1972), chap. 3; the latter is Patrick Thaddeus, in *Annual Reviews of Astronomy and Astrophysics*, vol. 10 (Annual Reviews, Inc., Palo Alto, California, 1972), p. 305.

The decisive measurement was reported by D. P. Woody, J. C. Mather, N. S. Nishioka, and P. L. Richards, *Physical Review Letters* **34**, 1036 (1975).

Appendix A

Historical Background and Sketch of the BCS Theory

As has been indicated in Chapters 4 and 14, the earliest theories of superconductivity were entirely phenomenological. Even before the discovery of the Meissner-Ochsenfeld effect, there had been some applications of thermodynamics to superconductivity, relating various thermal properties to the critical temperature T_c; but even some of those who did such work were suspicious of its validity. The first "explicative" proposal, made in 1934 jointly by C. J. Gorter of the Teyler Foundation, Haarlem, and H. B. G. Casimir of the University of Leiden, was that the electrons in a metal formed two "fluid" components, one of which behaved normally, while the other acted as if it were condensed into a "superfluid" interpenetrating the fluid of normal electrons. The fraction in the superfluid state is given by a quantity later designated the *order parameter*, ω, which varied from $\omega = 0$ at $T = T_c$ to $\omega = 1$ at $T = 0$.[1] This idea could account for many of the thermodynamic properties. Next was a theory proposed in 1935 by two brothers, Fritz and Heinz London, of the Clarendon Laboratory at Oxford University, who took the diamagnetic aspects as basic; they were able to "explain" the Meissner-Ochsenfeld effect, as well as the infinite conductivity, but with a penetration depth too small by one to two orders of magnitude. The Londons also recognized that quantum theory would be essential to a full understanding of superconductivity. In particular, they proposed that the diamagnetic properties were possibly the result of a "rigidity," or coherence, in the wave function of the electrons, which would give rise to currents in the presence of a magnetic field. Fritz London was especially active in the development of this concept, and in 1950 (by then at Duke University) he made it more explicit by suggesting that the electrons underwent a sort of "condensation," though in momentum space rather than in position, that would give rise to a gap or separation in energy between the ground state and the excited states.

An important experimental step was taken in 1950: Emanuel Maxwell of the National Bureau of Standards, and (independently) C. A. Reynolds, Bernhard Serin, W. H. Wright, and L. B. Nesbitt of Rutgers University found that the values of T_c for different isotopes of a single element (mercury) were different, varying approximately as the inverse square root of the atomic mass. This implied that the phenomenon of superconductivity arises from some sort of interaction between the electrons and the vibrations of the ionic lattice. Indeed, in the same year Herbert Fröhlich of the University of Liverpool independently proposed a theory along just such lines, as did John Bardeen a year later, but both attempts were essentially unsuccessful.

The mechanism for this type of interaction turned out eventually to be the correct one and so warrants some discussion. The electrons interact with each other directly, through the Coulomb potential, and also indirectly, by way of the ions in the lattice. The Coulomb interaction can be described as resulting from the exchange of virtual quanta of electromagnetic radiation, photons[2]; similarly, interaction via the lattice results from the exchange of quanta of lattice vibrational energy, called *phonons*. The resulting behavior depends in part on the relative strengths of these two interactions. As Fröhlich and Bardeen recognized, the second one can be attractive and can even be strong enough to overcome the Coulomb repulsion. Their difficulty lay in finding the proper way to utilize this result.

In the period from 1950 to 1954, A. B. Pippard of Cambridge University's Royal Society Mond Laboratory returned to the idea of coherence of the wave function and incorporated it into a generalization of the Londons' theory. With a coherence distance of about 10^{-4} cm, he was able to account for the actual values of the penetration depth in the Meissner-Ochsenfeld effect. Also in 1950, V. L. Ginzburg and L. D. Landau of the Institute for Physical Problems, Moscow, generalized the Londons' theory in another way, by regarding the order parameter ω as the square of the absolute value of a wave function Ψ—a concept which will be discussed later.

Ginzburg and Landau were able to set up equations for Ψ, which are especially useful when ω varies with position. The coherence length ξ appears naturally in their theory, in the form of a parameter which is essentially the inverse of the ratio of ξ to the Meissner-Ochsenfeld penetration depth.

The concept of a gap in the energy spectrum had been suggested by the Londons as early as 1935, but the first clear evidence for it came from experimental work by Heinz London in 1940. Shortly after Pippard's work on coherence, Bardeen showed that Pippard's form of the theory would be likely to follow from a model including such a gap. The idea of a gap received strong support from experiments between 1950 and 1958 on thermal conductivity, electronic contribution to specific heat, surface resistivity, and microwave and infrared transmission and absorption.

Both Fröhlich and Bardeen had tried to solve their model problem as an essentially one-body problem, treating the phonon-mediated interaction as a perturbation. However, the superconducting state is too drastically different from the normal state for this to be possible, as was proved in 1951 by M. R. Schafroth of the University of Sydney. In 1955, Schafroth showed that a gas composed of charged bosons—particles obeying Bose-Einstein statistics—would condense at a suitably low temperature into a superfluid such as was envisioned in the Gorter-Casimir theory; and he and his colleagues, John M. Blatt and Stephen T. Butler, attempted to construct a first-principles theory of superconductivity on this basis. They took into consideration the two-particle correlations brought about by the phonon-mediated interaction and tried to show that the pairs would indeed behave like charged bosons (they would not quite be truly bound states). They were able to achieve some measure of success only by assuming that the pairs were well localized in space, that is, separated by distances larger than the spatial extent of a given pair. They thus could not account for the relatively long-range, many-particle correlations that provide the coherence of the superconductive state.

The real breakthrough came when Cooper, in 1956, showed that for a sufficiently strong and attractive lattice-mediated force, there is indeed a strong correlation between pairs of electrons of opposite spins; but it is essentially a correlation in momentum "space" rather than in position. It is strongest when the two electrons of a pair have equal and opposite momenta; it decreases precipitously with increasing inequality. What Bardeen, Cooper, and Schrieffer did was to consider just the part of the energy arising from this correlation, on the basis that the rest would be the same in both the normal and the superconducting states. They then established that the state of lowest energy (at zero temperature) is the one in which all the electrons are paired off in this way, each pair having zero total momentum. The strong correlation in momentum implies, through the uncertainty principle, a large extent in space. Consequently, in contrast to the situation envisioned in the Blatt-Butler-Schafroth theory, a huge number of pairs (of the order of a million) have their centers of mass in the spatial volume occupied by a single pair.[3] It is this that produces the sort of coherence proposed by the Londons and Pippard; and the extent turns out to be, indeed, approximately 10^{-4} cm. The coherence is so strong that if a pair has a net momentum differing even slightly from that of the rest of the pairs, it dissociates and becomes part of the "normal" electron fluid. However, energy is needed to overcome the correlation and break up a pair, and it is this energy that gives rise to the gap in the energy spectrum.

It should be noted that the interaction of the electrons with the lattice, which is here presented as the mechanism responsible for superconductivity, is what gives rise to ordinary electrical resistance. A metal which is a good conductor in the ordinary sense, then, is one in which the electron-lattice interaction is weak, whereas superconductivity occurs only if it is sufficiently strong. Consequently, very good conductors such as the noble metals and the alkali metals do *not* become superconducting.

A key parameter in the BCS theory is an energy Δ that appears in the evaluation of the energy of the superconducting ground state in the case where the electrons form a homogeneous "gas." If the phonon-mediated coupling is not too strong—the precise limit is immaterial for the present discussion, as it is satisfied for most superconducting metals—Δ has the value $1.75k_B T_c$, where k_B is Boltzmann's constant. The difference in energy between the superconducting ground state and the lowest normal state, at zero temperature, then turns out to have the value $\frac{1}{2}N(0)\Delta^2$, where $N(0)$ is the number of states per unit energy interval at the Fermi energy level. This is actually a very small value, the energy per particle corresponding to roughly 10 millikelvins. Of more significance is the fact that Δ is the minimum energy per particle that must be provided to break a pair, so that the energy gap is of width 2Δ.

If the electrons are subject to an external potential and/or a magnetic field, then Δ also varies with position, and assumes the role of an effective potential acting on each particle of a pair. A computation of the energy for the case where the variation of Δ is not too rapid gives a result exactly parallel to that obtained in the Ginzburg-Landau theory; in particular, the Ginzburg-Landau form results if Δ is taken simply proportional to the Ginzburg-Landau wave function Ψ. Moreover, it is found that the charge and current densities are just those that would be obtained for particles of charge 2e described quantum mechanically by the wave function Ψ.[4] Thus Ψ is in a sense the wave function of the pairs. It is convenient to phrase the rest of the discussion in terms of it.

The function Ψ depends on position and time and may be complex (i.e., have both real and imaginary parts). Any such function can be written in the form

$$\Psi(r, t) = [\rho(r, t)]^{\frac{1}{2}} \exp[i\theta(r, t)],$$

where both ρ and θ are purely real and ρ is everywhere positive. The amplitude function ρ is related to the electric charge density; specifically, the charge density is just $2e\rho$. Both the amplitude function and the phase function θ—more properly, the spatial variation of θ—enter into the expression for the current density, as does the vector potential $\mathbf{A}$ that describes the magnetic field. The precise form of the relationship is not important here. What is important is that together with Maxwell's equations of electromagnetism, it leads to an equation for $\mathbf{A}$ in the interior of a semiconductor, whose solution is proportional to $\exp(-x/\lambda)$, where x is the distance from the surface and λ is a constant with the value approximately 2×10^{-5} cm. Thus $\mathbf{A}$ and with it the magnetic field drop to negligible values within a small fraction of a centimeter of the surface. This is, of course, the Meissner-Ochsenfeld effect, and λ is the penetration depth.

The fact that a spatial variation of θ is related to a current of pairs is also basic to the phenomena described in Chapter 14.

The time rate of change of θ consists of a linear term proportional to the total energy of the pair (including specifically a term proportional to the electrostatic potential V) together with a purely quantum mechanical term depending on the space and time variation of ρ. This last term may have a substantial effect at a boundary but is essentially zero in the interior of a semiconductor (because the charge density is uniform). It is the other part of the time dependence of θ that accounts for the perfect conductivity: If there were nonzero resistance to the pair current, then the current would produce a potential difference V between two points; this would give a growing phase difference between the two points, $2eVt/\hbar$, which would represent a breakdown of the basic correlation.

The relationship between potential difference and time variation of θ is also important in Chapter 14.

One final feature is of special significance. This is that the pairs, unlike the single electrons, are not subject to the exclusion principle. Rather, they behave in many respects like particles subject to Bose-Einstein statistics. In particular, not only is it possible for many of them to be in the same state; but the more of them there are in a given state, the greater is the probability that if additional ones are added to the system, they will enter that state rather than another. It is this that makes the BCS ground state so favorable and gives it its great coherence. It also emphasizes the many-body character of the BCS theory, and provides at least a basis in plausibility for the description of charge and current in terms of the pair wave function Ψ.

Notes

1. Compare the ideas of London regarding liquid helium II; see Chap. 5, p. 79.

2. A crude but useful classical analogy is the case of two people on a very smooth surface, such as ice, throwing a ball back and forth. This involves an interchange of energy and momentum between the people and gives rise to a repulsive interaction.

3. The presence of a very large number of pairs in a coherence volume is a source of difficulties for the Blatt-Butler-Schafroth formulation, as it means that correlation between not just two particles but all even numbers of particles must be included.

4. In discussions of superconductivity, the terms "order parameter" or "gap parameter" are usually used rather than "wave function."

Bibliography

Further descriptions of the BCS theory may be found in Leon N. Cooper, *American Journal of Physics* 28, 91 (1960), and in Richard P. Feynman, R. B. Leighton, and Matthew Sands, *The Feynman Lectures in Physics*, vol. III (Addison-Wesley, Reading, Mass., 1965), chap. 21; the latter includes a discussion of the Josephson effect. See also the Nobel lectures of J. R. Schrieffer, *Physics Today* 26, No. 7, 23 (1973), and John Bardeen, ibid., p. 41.

Name Index

Page numbers followed by n refer to notes; by b, to bibliography.

Subject Index

Page numbers followed by *n* refer to notes; by *b*, to bibliography.